QUICK START TO PROGRAMMING ALTERNATIVE CONTROLLOGIX® LANGUAGES

by

Jon Stenerson

Fox Valley Technical College – Appleton, WI

DELMAR
CENGAGE Learning™

Australia • Brazil • Japan • Korea • Mexico • Singapore • Spain • United Kingdom • United States

DELMAR
CENGAGE Learning

Quick Start to Programming Alternative ControlLogix® Languages
Jon Stenerson

Vice President, Career and Professional Editorial: Dave Garza

Director of Learning Solutions: Sandy Clark

Acquisitions Editor: Stacy Masucci

Managing Editor: Larry Main

Senior Product Manager: John Fisher

Editorial Assistant: Andrea Timpano

Vice President, Career and Professional Marketing: Jennifer Baker

Marketing Director: Deborah Yarnell

Marketing Manager: Katie Hall

Marketing Coordinator: Shanna Gibbs

Production Director: Wendy Troeger

Production Manager: Mark Bernard

Content Project Manager: Jim Zayicek

Art Director: David Arsenault

Technology Project Manager: Joe Pliss

For product information and technology assistance, contact us at
Cengage Learning Customer & Sales Support, 1-800-354-9706
For permission to use material from this text or product,
submit all requests online at **www.cengage.com/permissions.**
Further permissions questions can be emailed to
permissionrequest@cengage.com

Library of Congress Control Number: 2011921410

ISBN-13: 978-1-111-30971-8

ISBN-10: 1-111-30971-X

Delmar
5 Maxwell Drive
Clifton Park, NY 12065-2919
USA

Cengage Learning is a leading provider of customized learning solutions with office locations around the globe, including Singapore, the United Kingdom, Australia, Mexico, Brazil, and Japan. Locate your local office at: **international. cengage.com/region**

Cengage Learning products are represented in Canada by Nelson Education, Ltd.

To learn more about Delmar, visit **www.cengage.com/delmar**

Purchase any of our products at your local college store or at our preferred online store **www.cengagebrain.com**

Notice to the Reader

Printed in the United States of America
1 2 3 4 5 6 7 15 14 13 12 11

ControlLogix® is a registered trademark of Rockwell Automation Inc.

TABLE OF CONTENTS

PREFACE

I have been surprised at how quickly the new languages specified by IEC 11631-3 are gaining in popularity. Once the new languages are learned, they can dramatically reduce the time needed to program applications. The correct choice and use of language can simplify programming and also make the logic easier to understand and thus troubleshoot.

This book is for those who already understand PLC fundamentals and how to program in ladder logic. The book covers project organization and tag addressing, structured text programming, sequential function chart programming, function block programming, and Add-On instructions.

Chapter 1 concentrates on project organization and tag addressing. You should understand how projects are organized into tasks, programs, and routines. There is tremendous flexibility and power in the way projects are organized. The programmer has total control over when and how tasks, programs, and routines execute. Tag addressing, including array-type tags and user-defined tags, is also covered.

Chapter 2 examines the structured text language. Anyone who has done some computer programming will love structured text. It is very similar to C++, Pascal, Visual Basic, and other computer languages. Many types of industrial devices use languages that are very similar to structured text. Learning structured text programming will make all of these devices easy to learn and program. Structured text is also used in sequential function chart logic. I would suggest you study the chapter and then do the questions at the end before trying to program and test any logic.

Sequential function chart programming is covered in Chapter 3. Sequential function chart resembles a decision tree or a flow diagram. It is a very graphical language and is well suited to sequential type applications. It is very easy to understand a process when looking at a sequential function chart program.

Chapter 4 covers function block programming. Function block programming has many advantages. It is widely used on process control. The logic is easy to understand. There are many function block instructions available. The programmer chooses function blocks to do specific purposes and "wires" them together. Outputs from one function block instruction can become inputs to other function block instructions. Function block is one of the easier languages to learn.

Chapter 5 covers Add-On instructions. Add-On instructions are very powerful for the programmer. They can dramatically reduce repetitive work. They can also be used to simplify logic. Add-On instructions can also be used to protect proprietary logic.

In addition, there are tutorial/instructional videos available for download from a Student Companion site at www.cengagebrain.com.

HOW TO ACCESS A STUDENT COMPANION SITE FROM CENGAGEBRAIN

1. Go to http://www.cengagebrain.com
2. Type author, title or ISBN in the **Search** window
3. Locate the desired product and click on the title
4. When you arrive at the Product Page, click on the **Access Now** tab
5. Click on the Student Resources link in the left navigation pane to access the resources

INSTRUCTOR COMPANION WEBSITE

An Instructor Companion Website is available on www.cengagebrain.com. It contains an Instructor Guide providing answers to the end-of-chapter questions, chapter presentations done in PowerPoint, and additional chapter questions (ISBN 1-111-30971-X).
Instructor Set Up:

1. Go to http://login.cengage.com
2. If you already have a Cengage Learning Faculty Account, log in as a Returning User (even if it is your first time) and search for this book title and add it to your bookshelf.

ACKNOWLEDGMENTS

We would like to express appreciation to the following people for their input as reviewers of the first 4 chapters of this edition:

David Barth, Edison Community College, Piqua, OH
Charles Knox, University of Wisconsin, Platteville, WI
Wade Wittmus, Lakeshore Technical College, Cleveland, WI

CHAPTER

1

Memory and Project Organization

OBJECTIVES

Upon completion of this chapter, the reader will be able to:

- Describe project organization in ControlLogix (CLX).
- Explain the relationship between tasks, programs, and routines.
- List the types of task execution that are possible.
- Describe the base types of tags.
- Create base-, alias-, array-, and User-Defined-type tags.
- Choose the appropriate type of task execution and configure tasks.

INTRODUCTION

ControlLogix (CLX) was designed to give the programmer a great deal of flexibility in how an application is organized. CLX allows the programmer to keep things simple and program everything as one task or divide it into multiple tasks for efficient operation, clarity, and ease of understanding. Doing so gives tremendous flexibility and capability to the programmer. Therefore, it is very important to have a good understanding of CLX project organization and terminology.

CONTROLLOGIX PROJECTS

Typically, in most PLCs, we would have a program and maybe some subprograms to control an application. CLX has a different organizational model. The overall application that you develop in CLX is called a project. A project contains all of an application's elements and is broken into tasks, programs, and routines.

Tasks

A CLX project can have one or more tasks. Tasks can be used to divide an application (project) into logical parts. Tasks have a couple of important functions. A task is used to schedule the execution of programs in the task. A CLX project can have up to 32 tasks. A task's execution can be configured to be executed continuously, periodically, or on the basis of an event (see Figure 1-1). When the programmer creates a new project, a main task, which is continuous, is created. Continuous tasks are sometimes called the background tasks since they execute only in leftover time. The name *main task* is somewhat misleading. It is actually the lowest priority task. It can be renamed.

Task Execution	Function
Continuous	Operates continuously (except while other tasks are executing)

Periodic	Executes at specific intervals.
	    The rate of execution can be set between 1 ms to 2000 seconds.
	    The default execution time is 10 ms.

Event-based	Executes on the basis of an event

Figure 1-1   Task execution types.

A continuous task can be thought of as executing continuously. As only one task can execute at a time, a continuous task executes anytime a periodic or event-based task is not executing. Periodic tasks are set up to operate one time though at specified intervals. Periodic tasks interrupt the operation of the continuous task. When the periodic task is done, the task is executed one more time. The rate for a periodic task can be set to lie between 0.1 ms and 2000 s. The default rate for a periodic task is 10 ms.

Figure 1-2 shows a timing diagram for three tasks. The main task is continuous. It is shown in gray. It is always operating if the other two are not. Task 2 (white) is a periodic task. It executes at specific time intervals. The main task (continuous) stops executing and the second task (periodic) executes. The third task (black) is event based. It executes when the specified event occurs. Remember that only one task can execute at a time.

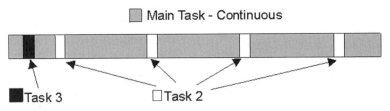

Figure 1-2   Continuous, periodic, and event-based task execution.

Application Example

Imagine a machine that produces packaging material. The machine requires several servosystems for motion, velocity, flow control, temperature control, and many quality control checks as the packaging material is made. This machine application might be broken into several tasks for a CLX project. The main task might be used for overall machine control functions. The company also collects machine production data and displays it for operators on human–machine interface (HMI) monitors. The main task is a continuous task. In this example, the servo motion and process control need to be monitored for safety and for adequate control. This needs to be done in a periodic task. Another operation on this machine is making a perforation. This must occur on the basis of a registration mark on the packaging material. This task would require event-based execution. In this example, the project developer might decide to divide the overall application (CLX project) into three tasks, as each has different requirements. Figure 1-3 shows what the project organization might look like for this application. This application (project) was broken into three tasks. One task is continuous, one needs to execute about every 5 ms (periodic), and one is based on the registration mark on the packaging material (event based). Tasks will be covered in greater detail later in this chapter.

Programs

As Figure 1-3 shows, a project consists of all of the things required to control an application. A separate task can be developed to control each logical portion of an application.

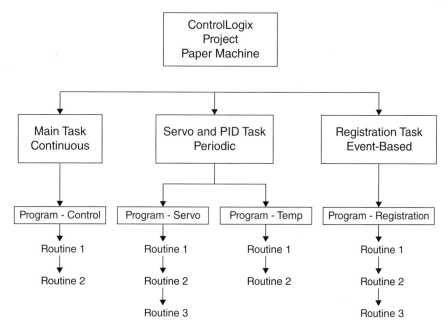

Figure 1-3 ControlLogix project organization with the application organized into tasks, programs, and routines.

CLX also enables the programmer to break each task into one or more programs. Each task can have up to 100 programs. In Figure 1-3, the task named Main Task has one program named Control: The tasks named Servo and PID have two programs: Servo and Temp. The third task, named Registration, has one program. If there is more than one program, the programs will execute in the order they are shown in the controller organizer.

Routines

Each program can also have one or more routines. An application's logic is created in the routines. These are normally organized into a main routine and additional subroutines. In most PLCs, the logic is written in programs and subprograms. In CLX, they logic is written in routines.

Figure 1-4 shows a different representation of project organization. A CLX project does not have to be complex. As shown in Figure 1-4, a simple project might just have one task, one program, and one routine. There is one task named MainTask, one program named MainProgram, and one routine named MainRoutine.

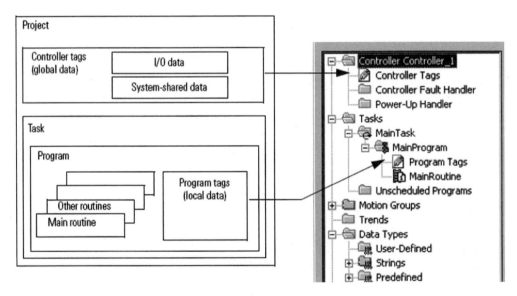

Figure 1-4   Project organization.

LET'S REVIEW

Project

A project is the overall complete application. It is the file that stores the logic, configuration, data, and documentation for a controller.

Task

A task is a scheduling mechanism for executing programs. An application can be broken into multiple logical tasks. A task enables the programmer to schedule and prioritize one or more programs that execute on the basis of the application requirements.

When a new project is started, a continuous task is created by default. It is preconfigured as a continuous task. The programmer can add additional periodic or event tasks, as needed. Once a task is triggered, every program assigned (scheduled) to the task will execute in the order in which it is displayed in the controller organizer.

Program

In CLX a program has one or more routines. The routines contain the logic in a CLX project. A program could be defined as a set of related routines and tags. Each program contains program tags, a main executable routine, other routines, and an optional fault routine. The programmer can write logic for a fault routine. The fault routine runs in the event of a controller fault. Programs are contained in a task: When the program's task is triggered, the scheduled programs within the task will execute from the first to the last one.

Routine

A routine is a set of logic instructions written in one programming language, such as ladder logic. A routine in CLX is similar to a program or subprogram in most PLCs. Routines are where the programmer writes the executable code for the project.

The main routine is the first routine to execute when its program is triggered to run. Jump-to-subroutine (JSR) instructions are used to execute other routines. A program fault routine can also be developed. If any of the routines in the associated program produce a major fault, the controller executes the program fault routine, if one was developed.

TAG ADDRESSING IN CONTROLLOGIX

In most PLCs, the programmer has to use very specific addressing to specify I/O addresses, bits, variables, timers, counters, and so on. Most PLCs use a physical address for every tag. For example, a SLC 500 PLC would use an address like N7:5 to reference an integer in memory. Addresses typically follow a fixed, numeric format that depends on the type of data, such as B3:6/0, N7:2, and F8:5.

In these PLCs, the programmer can use symbolic names to represent the actual address. For example, the programmer might use a symbolic name such as Alarm_Light to represent an actual output (O:5/3) in a ladder diagram. The PLC actually uses the O:5/3 address. The symbolic name for it is not even located in PLC memory. It only appears in the program on the computer. It is only for the programmer's use.

In CLX, tags are used to address I/O, bits, variables, timers, counters, and so on. A tag is a user-friendly name for a memory location. For example, we might store a temperature integer value in memory. Temp would be a good name for the tag to hold this data. The processor uses the tag name to address the data.

CLX uses the tag name and doesn't need to cross-reference a physical address. The tag name identifies the data. This enables a programmer to document a program with tag names that clearly represent the application. In CLX the maximum length for a tag name is 40 characters.

Tag names may use alphabetic characters (A–Z or a–z), numeric characters (0–9), and underscores (_). Tag names must start with either an alphabetic character or an underscore. Tags are not case sensitive (A is the same as a). It is wise to use mixed-case tag names (upper- and lowercase characters) and underscores because mixed-case tag names are easier to read. Look at the examples in Figure 1-5.

Preferred Tag Name	More Difficult to Read
Temp_1	TEMP_1
Temp_1	TEMP_1
	temp_1

Figure 1-5   The use of uppercase and lowercase letters and underscores can make tag names easier to read.

Organizing Tags

RSLogix 5000 organizes tags in alphabetical order. Tag names can be chosen so that they keep similar data together. For example, if we are interested in tags related to Machine_1 or tags related to temperature, we can name them so that they are listed together.

The first column in the table in Figure 1-6 shows an example of naming tags so that similar tags are grouped together. All the tags related to Machine 1 appear together, as do the Temp tags. In the second column similar tags are not grouped together because of their names. They are separated from each other. One would have to go through the list to find each tag that is related to Machine 1 or Temp.

Tag Data Types

The data type could be defined as the type of data that a tag stores, such as a bit, an integer (whole number), a real (floating-point) number, a string, and so on. The minimum memory allocation for a tag is 4 bytes (32 bits) plus 40 bytes for the tag name itself. If a tag type that uses fewer than 4 bytes of memory is used, the controller allocates 4 bytes for it anyway. A BOOL-type tag, for example, requires only 1 bit, but the controller allocates 4 bytes to store it, 1 bit for the actual tag value and 31 unused bits.

Figure 1-7 shows the basic types of tags in a CLX project: base, alias, produced, and consumed.

Base-Type Tags
A base-type tag would usually be chosen to create tags that would hold data for logic. For example, we would choose base type for tags to hold temperature, quantities, bits,

Logical Organization	No Name Organization
Tag Name	Tag Name
Machine_1_Cyc	...
Machine_1_Hours	Coil_1_Temp
Machine_1_On	Tag names that are between C and E
Machine_1_Stat	...
...	...
...	...
Temp_Coil_1	...
Temp_Extruder	Extruder_Temp
Temp_Heater	
Temp_Machine_Ldr	

Figure 1-6   Tag names. The first column shows an example of careful tag naming so that similar tags are grouped together. In the second column, similar tags are not grouped together by the first word of their name.

Tag Type	Use of This Type of Tag
Base	Stores various types of values for use by logic in the project
Alias	Represents another tag
Produced	Sends data to a different controller
Consumed	Receives data from a different controller

Figure 1-7   Basic tag types.

integer numbers, and floating-point (real) numbers. Figure 1-8 shows some of the numerical types for base-type tags. Figure 1-9 shows the size number each base-type tag can hold.

The Boolean (BOOL)-type tag is one of the more commonly used tag types. The BOOL is a bit tag that can have a value of 1 or 0. It is 1 bit in length (see Figure 1-2), although it takes up 32 bits in memory.

A single-integer (SINT)-type tag is 8 bits in length, although it uses 32 bits in memory. This type of tag can hold a value between 2128 and 1127. A SINT tag is used for whole (nondecimal) numbers.

An integer (INT)-type tag can be used to hold a value between 232,768 and 132,767. An INT tag uses 16 bits to hold a value, but uses 32 bits in memory. An INT tag is used for whole numbers. One use of an INT-type tag is when we communicate

Type of Tag	Use
BOOL	Bit
BOOL	Digital I/O points
CONTROL	Sequencers
COUNTER	Counter
DINT	Integer (whole number, 32 bit)
INT	Analog device in integer mode (very fast sample rates)
REAL	Floating-point (decimal) number
TIMER	Timer

Figure 1-8   Number types for base-type tags.

Type	Bit Use and Size of Numbers for Each Type						
	31	16	15	8	7	1	0
BOOL							0 or 1
SINT					−128 to +127		
INT			−32,768 to +32,767				
DINT	−2,147,483,648 to +2,147,483,647						
REAL	−3.40282347E^{38} to −1.17549435E^{-38} (negative values) 0 1.17549435E^{-38} to 3.40282347E^{38} (positive values)						

Figure 1-9   Size of numbers that each base type can hold.

between a CLX controller and a SLC. The length of an integer in a SLC is the same as an INT in a CLX controller.

A double-integer (DINT)-type tag is used for whole numbers. A DINT tag uses 32 bits to hold a value. A DINT can hold a value between 22,147,483,648 and 12,147,483,647.

A REAL-type tag is used to hold decimal (floating-point) values. A REAL tag is also 32 bits.

Alias-Type Tags
An alias-type tag is used to create an alternative name (alias) for a tag. The alias tag is often used to create a tag name to represent a real-world input or output. An alias is indeed a tag unto itself, not just another name for the base tag. It is linked to the base tag so that any action to the base also happens to the alias and vice versa. Figure 1-10 shows an

example of the use of alias tags. Note that the alias tag name and the actual address (the base tag) of the input and output are shown in the rung.

Figure 1-10   Rung showing a contact and coil. Note the alias name (Fan_Motor) and the base tag (<Local:2:0.Data.5>) of the output coil. The alias name is easier to understand and easier to relate to the application, although the base tag contains the physical location of the output point in the ControlLogix chassis.

Figure 1-11 shows how a tag is configured in CLX. The name is entered first. The tag type is chosen: Base, Alias, Produced, or Consumed. Next the Data Type is chosen. The Tag Properties screen also allows the programmer to choose the Scope of and the Style in which to display the tag.

Figure 1-11   How a tag is configured in CLX. Note the choices: Base, Alias, Produced, and Consumed. Note that the programmer also has a choice of the Scope and Style of the tag in this screen.

Scope of Tags

Scope refers to which programs have access to a tag. There are two scopes for tags in CLX: controller scope and program scope (see Figure 1-12). A controller scope tag is available to every program in the project. The controller scope tag data is also available to the outside world, such as SCADA systems. Program scope tags are available only within the program they are created in. Programs cannot access or use a different program's tags if the tags were created to be program scope. Figure 1-12 shows two programs within a project (Program A and Program B). Note that each program has tags named Tag_1, Tag_2, and Tag_3. Note that the names of tags are the same in both programs. They are not, however, the same tags. There is no relationship between them, even though they have the same name. They are program scope tags. They are available only to routines within that program. Note, however, that there are some controller scope tags: Sensor_1, Temp_1, and CNT. These are available to all programs because they were created as controller scope tags. This means, for example, that Temp_1 is available to both programs and is the same tag for both programs.

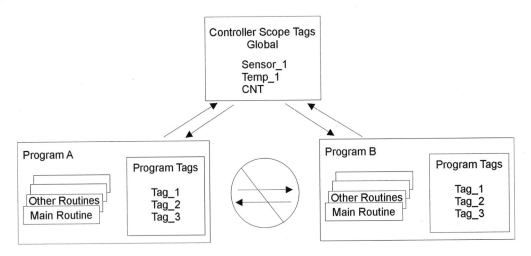

Figure 1-12   Scope of tags.

Routines within a program can access program tags of only the program they belong to and controller tags. Routines cannot access the program tags of other programs.

Creating a Tag

There is more than one way to create tags. Tags can be created one at a time as you program or tags can be created in the tag editor. The tag editor enables you to create and edit tags using a spreadsheet-like view of the tags. Figure 1-13 shows the tag editor screen with three tags that have been created. Note that these three tags are program scope tags. The name of the program is MainProgram, and program scope is the scope that was chosen.

Note the **Edit Tags** tab was chosen on the bottom of the screen. You must be in the edit mode to add or edit tags. The monitor mode is used to monitor tag values.

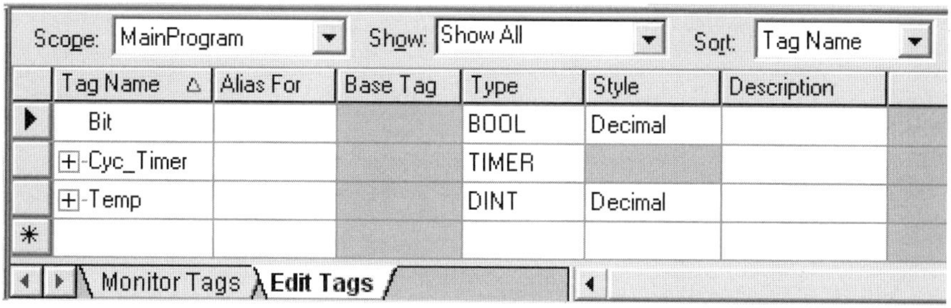

Scope: MainProgram ▼		Show: Show All ▼		Sort: Tag Name ▼	
Tag Name △	Alias For	Base Tag	Type	Style	Description
▶ Bit			BOOL	Decimal	
⊞-Cyc_Timer			TIMER		
⊞-Temp			DINT	Decimal	
*					

\\ Monitor Tags \ **Edit Tags** /

Figure 1-13 Tag editor screen. Note that three tags have been defined. Bit is a BOOL-type tag. Cyc_Timer is a TIMER-type tag. Temp is a DINT-type tag. Note the scope of these tags is program scope (named MainProgram). Note also that the two selection tabs at the bottom are Monitor Tags and **Edit Tags** and that the **Edit Tags** tab is active.

Arrays

Logix5000 controllers also allow you to use arrays to organize data. Arrays are very important in CLX programming. An array is a tag type that contains a block of multiple pieces of data. An array is similar to a table of values. Within an array of data values, each individual piece of data is called an element. Each element of an array must be of the same data type. An array tag occupies a contiguous block of memory in the controller; each element is in order. Arrays are useful if you want to index (move) through the elements of an array. Arrays can be created with 1, 2, or 3 dimensions.

An array is like a table of tags (see Figure 1-14). It can hold the values of multiple tags. For example, an application might require five different temperatures, one for each different product that is produced. Figure 1-14 shows an example of a 1-dimensional array created to hold 5 temperatures. The tag name is Temp. A subscript identifies each individual element within the array.

Temp[0]	210
Temp[1]	200
Temp[2]	190
Temp[3]	180
Temp[4]	170

Figure 1-14 This is a 1-dimensional (one column of values) array. There are five members of this array, Temp[0] to Temp[4]. Each member of the array has a different value in this example.

Figure 1-15 shows a 2-dimensional array. Note that there are 3 columns and 5 rows in this example. This would be a 2-dimensional 5 by 3 array and would have 15 members.

Temp[0,0]	225	Temp[0,1]	200	Temp[0,2]	175
Temp[1,0]	220	Temp[1,1]	195	Temp[1,2]	170
Temp[2,0]	215	Temp[2,1]	190	Temp[2,2]	165
Temp[3,0]	210	Temp[3,1]	185	Temp[3,2]	160
Temp[4,0]	205	Temp[4,1]	180	Temp[4,2]	155

Figure 1-15   A 2-dimensional array.

Arrays can also have 3 dimensions. This would be like a cube of values.

Figure 1-16 shows a 1-dimensional array of 10 values. The tag name of the array is Temp. There are 10 members: Temp[0]–Temp[9]. The values of each member are shown in the second column. The fourth column shows that they are all of type DINT. Remember that arrays can be created for any type of data but an array can hold only one data type.

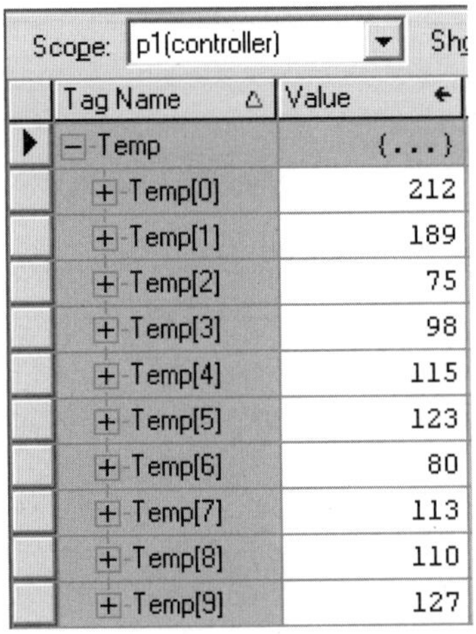

Scope: p1(controller) ▼	Sho
Tag Name △	Value ←
▶	─ Temp	{...}
+ Temp[0]	212
+ Temp[1]	189
+ Temp[2]	75
+ Temp[3]	98
+ Temp[4]	115
+ Temp[5]	123
+ Temp[6]	80
+ Temp[7]	113
+ Temp[8]	110
+ Temp[9]	127

Figure 1-16   An array of 10 DINT-type tags.

Creating Arrays

To create an array, you create a tag and assign dimensions to the data type. From the Logic menu, select Edit Tags. Type a Name for the tag and select a Scope for the tag (see left side of Figure 1-17). Assign the Array Dimensions (see right side of Figure 1-17). In this example, it will be a 1-dimensional array. There will be 5 values. Note that in Array Dimensions 5 was entered in Dim 0, and 0 was left in Dim 1. Figure 1-18 shows how the Temp array would appear in the CLX tag editor. Note the

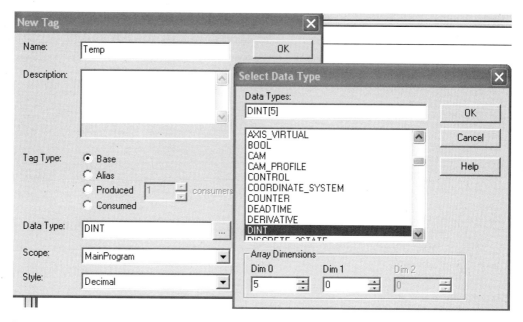

Figure 1-17   Creating an array of 5 DINTs.

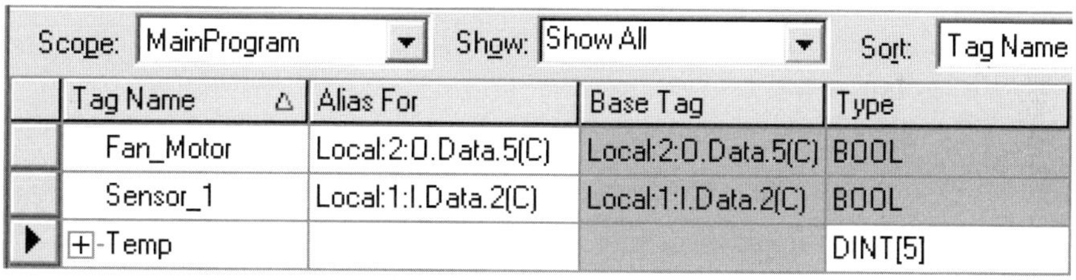

Figure 1-18   Tag editor showing the Temp array.

1 to the left of Temp. If you click on the 1, the five members of the array will appear. Figure 1-19 shows the five members. Note that there is a 1 to the left of each member. You can click on the 1 to see each bit within the member. There are 32 bits for each array member (DINT).

Produced/Consumed Tags

If we would like to share tag information with multiple controllers, produced- and consumed-type tags can be used. If we wanted to make a tag available from PLC 1 to PLC 2, it would be a produced-type tag in PLC 1 and a consumed-type tag in PLC 2.

Scope:	MainProgram ▼	Show:	Show All ▼	Sort:	Tag Name ▼

Tag Name △	Value ←	Force Mask ←	Style	Type
Fan_Motor	0		Decimal	BOOL
Sensor_1	0		Decimal	BOOL
─ Temp	{ . . . }	{ . . . }	Decimal	DINT[5]
+ Temp[0]	0		Decimal	DINT
+ Temp[1]	0		Decimal	DINT
+ Temp[2]	0		Decimal	DINT
+ Temp[3]	0		Decimal	DINT
+ Temp[4]	0		Decimal	DINT

Figure 1-19 Tag editor showing an array-type tag that was created. Note there are 5 members in this array: Temp[0] through Temp[4].

Structures

Remember that arrays can hold only one data type. CLX offers another type of tag that can hold multiple types of data. This type is a structure. Structures enable the programmer to create a structure-type tag that can hold multiple data types.

A structure can be created to match a specific application's requirements. Each individual data type in a structure is called a member. Members of a structure have a name and data type, just like tags. CLX has several predefined structures (data types) for use with specific instructions, such as timers, counters, motion instructions, function block instructions, and so on. Users can create their own structure tags, called a User-Defined data type.

Figure 1-20 shows an example of a structure-type tag. A tag named TMR_1 was created. The type chosen for it was TIMER. Once you have created the tag and tag type for a timer, CLX creates the tag and 9 additional tag members. Note in the figure that the first member shown is TMR_1.PRE. This is the preset value for this timer tag. Note the period between the name of the timer and the tag member (PRE in this example). Note also that PRE is a DINT type. The next tag member is ACC. It is also a DINT type. Next, look at the DN member. This would be set to a 1 when the accumulated (ACC) value is equal to the preset (PRE) value. The DN member is a BOOL type. A structure-type tag can hold several pieces of data and each member can be of a different type. Structures are also used for counters, motion, and many other purposes in CLX.

User-Defined Structure Tags

Programmers can create their own structure type for tags. These are called User Defined. An example is shown in Figure 1-21. To create a User-Defined structure

TMR_1	{...}	{...}		TIMER
TMR_1.PRE	30000		Decimal	DINT
TMR_1.ACC	0		Decimal	DINT
TMR_1.EN	0		Decimal	BOOL
TMR_1.TT	0		Decimal	BOOL
TMR_1.DN	0		Decimal	BOOL
TMR_1.FS	0		Decimal	BOOL
TMR_1.LS	0		Decimal	BOOL
TMR_1.OV	0		Decimal	BOOL
TMR_1.ER	0		Decimal	BOOL

Figure 1-20   A structure-type tag for a timer named TMR_1. Note there are 9 members: PRE, ACC, EN, TT, and so on. Note also that PRE and ACC are DINT types and the rest of the members are BOOLs.

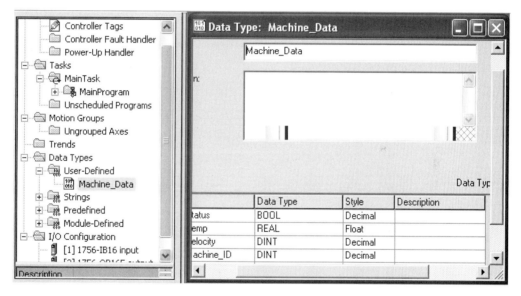

Figure 1-21   A User-Defined structure-type tag. Note that the new tag type is named Machine_Data and there are 4 members: Status (BOOL), Temp (REAL), Velocity (DINT), and Machine_ID (DINT).

tag, the programmer must right click on **User Defined** in the Controller organizer and choose **New Data Type** and then enter the tag name, members, and their types. This tag will then be available as a new tag type for programming. The programmer has many machines in his or her company. The machines are all quite similar and have mostly the same type of data available. The programmer created the new User-Defined-type tag to combine the common data for each machine and ease programming. In Figure 1-21 the name of the new User-Defined type is Machine_Data. Note that four tag names as well as the type for each tag were entered. These will be the tag members in the new User-Defined structure tag named Machine_Data.

This new tag type is now available and can be used when new tags are created.

Once the User-Defined-type tag has been created, the programmer can use the new tag type. In Figure 1-22 the programmer created a tag named Weld_Machine in the tag editor. The programmer then chose Machine_Data as the type. Remember that Machine_Data was just created as a User-Defined type. When the programmer created the new User-Defined type, it was automatically added to the Data Types choice list. Now when the programmer creates a tag named Weld_Machine, the software automatically creates the four tag members.

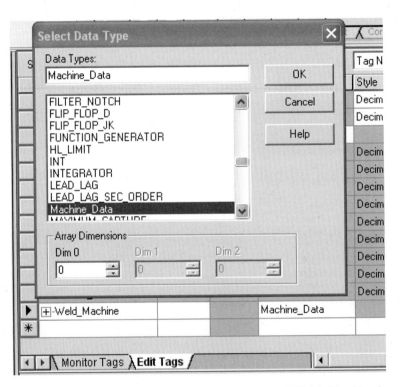

Figure 1-22 The programmer entered a new tag name (Weld_Machine) and chose Machine_Data as the type.

Figure 1-23 shows the new tag that was created (Weld_Machine) and its members. Note that the four members created were of the types specified in the Machine_Data User-Defined tag.

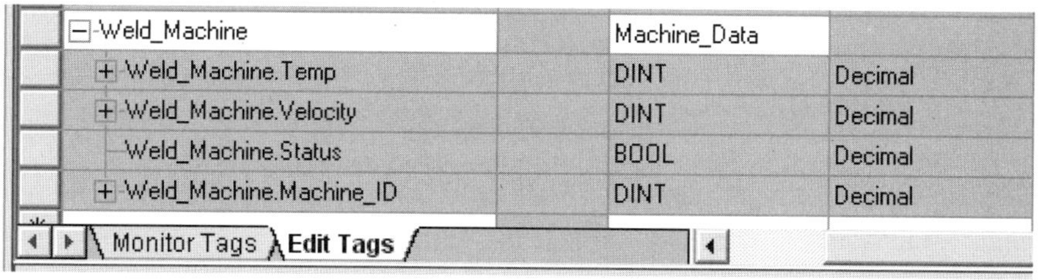

⊟ Weld_Machine	Machine_Data	
⊞ Weld_Machine.Temp	DINT	Decimal
⊞ Weld_Machine.Velocity	DINT	Decimal
Weld_Machine.Status	BOOL	Decimal
⊞ Weld_Machine.Machine_ID	DINT	Decimal

◄ ► \ Monitor Tags ↘ **Edit Tags** /   ◄

Figure 1-23   Tag members for the tag named Weld_Machine.

Guidelines for User-Defined Structure Data Types
When you create a User-Defined data type, keep the following in mind:

> If members that represent I/O devices are used, logic must be used to copy the data between the members in the structure and the corresponding I/O tags.
> If you include an array as a member, it can only be a 1-dimensional array. Multidimensional arrays cannot be used in a User-Defined data type.
> When you use the BOOL, SINT, or INT data types, put members that use the same data type in order. User-Defined data types can be used inside other User-Defined data types.

Figure 1-24 reviews the definition of *data type* and *structure*.

Term	Definition
Data type	The type of data a tag can store (BOOL, SINT, REAL, etc.)
Structure	A tag type that holds more than one tag and more than one type of tag.
	Each individual data type in a structure is called a member.
	Members each have a name and a data type.
	CLX has some standard structures available for counters, timers, etc.
	Users can create their own structures for specific uses. These are called User-Defined data types.

Figure 1-24   Definitions of data type and structure.

REAL-WORLD I/O ADDRESSING

Addressing of real-world I/O is different in CLX than in other PLCs. Study Figure 1-25. The first part of the address is the Location. This can be Local, meaning in the same chassis as the controller, or the name of a remote communications adapter or a communications bridge module. A colon follows the Location. The next part of the address

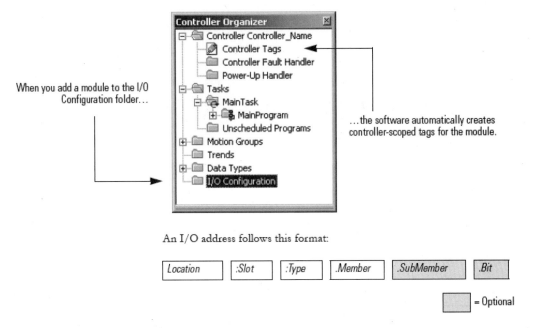

When you add a module to the I/O Configuration folder...

...the software automatically creates controller-scoped tags for the module.

An I/O address follows this format:

Location	:Slot	:Type	.Member	.SubMember	.Bit

☐ = Optional

Figure 1-25   Real-world tag addressing.

is the Slot number of the I/O module in the chassis. After another colon, the Type of data follows. This can be an I (input), a O (output), a C (configuration), or an S (status). A period delimiter is used next, followed by the Member. The Member specifies the specific data from the I/O module. If it is a digital input or an output module, it stores the bit values for the module. If it is an analog module, it stores the data for a channel. A period delimiter is next, followed by the Submember. The Submember is specific data related to a member. Another period follows and then the Bit. The Bit is a specific point on a digital module, one bit of an output module, for example. Figure 1-26 shows a table that explains each part of an address. Occasionally the Member or Submember does not exist.

The good news is that the programmer does not have to type the address for an alias tag. When an alias tag is created, the address can be chosen from the available controller I/O.

I/O Module Tags

When you add modules to a project, tags are automatically created for the modules. Figure 1-27 shows the Controller Organizer after an input module (slot 3) and an output module (slot 4) were added. Note that they were diagnostic modules. Diagnostic modules have more tags than many of the other I/O modules. RSLogix5000 automatically creates the correct controller scope tags for the modules that are installed.

Address	Content
Location	Network location Local = same chassis as controller Adapter_Name = name of a remote communications adapter or a bridge module
Slot	Slot number of I/O module in the chassis
Type	Type of data I = input O = output C = configuration S = status
Member	Specific data from the I/O module For a digital module, stores input or output bit values for the module For an analog module, stores the data for a channel
SubMember	Specific data related to a member
Bit	Specific point on a digital I/O module

Figure 1-26   CLX real-world tag addressing.

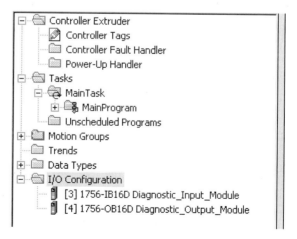

Figure 1-27   Two modules added to a project.

Figure 1-28 shows tags that were created for the input module that was added. Note that the top portion of the tag editor (Local:3:C) has Configuration tags. These are set when the module is configured. Configuration tags determine the characteristics and operation of a module.

Scope: Extruder(controller) ▼	Show: Show All ▼	Sort: Tag Name ▼
Tag Name △	Value	←
▶ ─ Local:3:C		{...}
Local:3:C.DiagCOSDisable		0
+ Local:3:C.FilterOffOn_0_7		1
+ Local:3:C.FilterOnOff_0_7		1
+ Local:3:C.FilterOffOn_8_15		1
+ Local:3:C.FilterOnOff_8_15		1
+ Local:3:C.COSOnOffEn	2#0000_0000_0000_0000_1111_1111_1111_1111	
+ Local:3:C.COSOffOnEn	2#0000_0000_0000_0000_1111_1111_1111_1111	
+ Local:3:C.FaultLatchEn	2#0000_0000_0000_0000_1111_1111_1111_1111	
+ Local:3:C.OpenWireEn	2#0000_0000_0000_0000_1111_1111_1111_1111	
─ Local:3:I		{...}
+ Local:3:I.Fault	2#0000_0000_0000_0000_0000_0000_0000_0000	
+ Local:3:I.Data	2#0000_0000_0000_0000_0000_0000_0000_0000	
+ Local:3:I.CSTTimestamp		{...}
+ Local:3:I.OpenWire	2#0000_0000_0000_0000_0000_0000_0000_0000	

Figure 1-28 Tag editor showing tags that were automatically added after the input module in slot 3 was added.

The bottom portion of the tag editor (Local:3:I) shows the input tags that are available. The first ones are Fault bits that can be used in logic or for troubleshooting. The second set, labeled Data, contains the actual input bits from the module. The third set of inputs is the CST Timestamp (CST stands for coordinated system time) information. The last set of inputs are the OpenWire inputs.

Figure 1-29 shows tags that were automatically created for the output module that was added. Note that the top portion of the tags (Local:4:I) are input-type tags. These include Fault, Data, CST Timestamp, FuseBlown, NoLoad, and OutputVerifyFault.

Next are the actual outputs (Local:4:O.DATA). This module has 16 outputs, given in the first 16 bits.

The last set of tags are Configuration tags. These are set when the module is configured. Configuration tags determine the characteristics and operation of a module.

Scope: Extruder(controller) ▼	Show: Show All ▼	Sort: Tag Name ▼

Tag Name △	Value
− Local:4:I	{...}
+ Local:4:I.Fault	2#0000_0000_0000_0000_0000_0000_0000_0000
+ Local:4:I.Data	2#0000_0000_0000_0000_0000_0000_0000_0000
+ Local:4:I.CSTTimestamp	{...}
+ Local:4:I.FuseBlown	2#0000_0000_0000_0000_0000_0000_0000_0000
+ Local:4:I.NoLoad	2#0000_0000_0000_0000_0000_0000_0000_0000
+ Local:4:I.OutputVerifyFault	2#0000_0000_0000_0000_0000_0000_0000_0000
− Local:4:O	{...}
+ Local:4:O.Data	2#0000_0000_0000_0000_0000_0000_0000_0000
▶ − Local:4:C	{...}
Local:4:C.ProgToFaultEn	0
+ Local:4:C.FaultMode	2#0000_0000_0000_0000_0000_0000_0000_0000
+ Local:4:C.FaultValue	2#0000_0000_0000_0000_0000_0000_0000_0000
+ Local:4:C.ProgMode	2#0000_0000_0000_0000_0000_0000_0000_0000
+ Local:4:C.ProgValue	2#0000_0000_0000_0000_0000_0000_0000_0000
+ Local:4:C.FaultLatchEn	2#0000_0000_0000_0000_1111_1111_1111_1111
+ Local:4:C.NoLoadEn	2#0000_0000_0000_0000_1111_1111_1111_1111
+ Local:4:C.OutputVerifyEn	2#0000_0000_0000_0000_1111_1111_1111_1111

Figure 1-29   Tag editor showing tags that were automatically added after the output module in slot 4 was added.

MORE ON THE USE OF TASKS

Remember that a CLX project can have multiple tasks and that tasks can be scheduled. The default RSLogix 5000 project provides a single task for all your logic. Although this is sufficient for many applications, some situations may require more than one task.

A Logix5000 controller supports multiple tasks that can be used to schedule and prioritize the execution of programs. This can help balance the processing time of the controller. Remember that

- A controller can execute only one task at one time.
- A task that is executing can be interrupted by another higher-priority task.
- Only one program executes at one time in a task.

Figure 1-30 explains the three possible task execution types and the characteristics of each type.

To Execute a Task	Use a(n)	Description
Continuously	Continuous task	Runs in the background. Any CPU time that is not allocated to other operations is used to execute the programs in the continuous task. Other operations would include CPU time for communications, motion control, and periodic or event-driven tasks.
		Runs all of the time. When a scan is complete a new one begins immediately.
		Projects do not require a continuous task. There can only be one continuous task.
At a periodic rate, multiple times within the scan of the other logic	Periodic task	Executes at a specific time period. When the time period occurs, a periodic task Interrupts any task with a lower priority and executes once Returns control to where the previous task left off The time period can be configured from 0.1 ms to 2000 s. Default time is 10 ms.
Immediately when an event occurs	Event task	Only executes when a specific event (trigger) occurs. When the event occurs, an event task Interrupts any lower-priority task, and executes once Returns control to where the previous task left off The trigger can be a(n) Digital input New sample of analog data Certain motion operation Consumed tag Event instruction

Figure 1-30 Task execution types.

It is important to choose the correct type of execution for each task. The table in Figure 1-31 shows examples of types of applications and which type of task execution they might be best suited to.

Application Example	Type of Task
Fill a tank and control its level (Without PID)	Continuous
Monitor, control, and display application parameters	
Monitor a tag every 0.1 s and calculate a rate of change to be used for control	Periodic
Perform quality measurements every 40 ms	
Control level with PID	
On a packaging line, seal the package immediately when a registration mark is sensed	Event
If a specific alarm is sensed, shut down the machine immediately	
When a box arrives at the taping position, execute the taping routine immediately	

Figure 1-31 Examples of task execution types.

Number of Tasks

A CLX CPU can support up to 32 tasks. Only one task executes at a time and only one task can be continuous. It is possible to have too many tasks. Every task takes controller time away from the other tasks when it executes. If there are too many tasks, it is possible that tasks may overlap. If a task is interrupted for too long or too frequently, it may not complete its execution before it is triggered again. It is then possible that the continuous task may take too long to complete.

At the end of a task's execution, the controller performs overhead operations (output processing) for the I/O modules in the system. The output processing may affect the update of the I/O modules in the system. Output processing can be turned off for a specific task; this reduces the elapsed time of that task.

Every task has a watchdog timer that specifies how long a task can execute before it triggers a major fault. It is assumed that something has gone wrong if a task takes too long to execute. The watchdog timer begins to accumulate time when the task is initiated and stops accumulating time when all the programs within the task have executed. A watchdog time can range from 1 ms to 2000 s. The default time is 500 ms.

If a task takes longer than the specified watchdog time, a major fault occurs. The time includes interruptions by other tasks. A watchdog time-out fault may also occur if a task is repeatedly triggered while it is still executing. This can occur if a lower-priority task is interrupted by a higher-priority task, and it will delay the completion of the lower-priority task.

It is possible to use the controller fault handler to clear a watchdog fault. However, if the same watchdog fault occurs again during the same logic scan, the controller enters faulted mode, regardless of whether the controller fault handler clears the watchdog fault.

Setting the Watchdog Time for a Task

The watchdog timer is a preset parameter that the programmer can configure. The watchdog timer monitors the scan time of a task. If the watchdog timer reaches the PRE value, a major fault occurs. Depending on the controller fault handler, the controller may shut down. To change the watchdog time of a task, right click the task and select **Properties**. Next select the **Configuration** tab and set the watchdog time-out for the task in milliseconds (see Figure 1-32). The watchdog default time of 500 ms is shown in Figure 1-32.

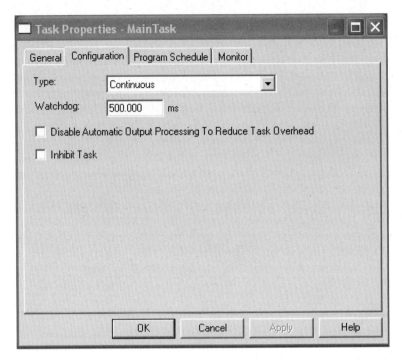

Figure 1-32 Setting the watchdog timer.

QUESTIONS

1. List the three main components that a project is composed of and describe each.
2. Which component in a project contains the logic?
3. What are the three main types of tasks?
4. What is a project?
5. What is the minimum memory allocation for a tag?
6. What is a base-type tag?
7. What is a DINT?
8. What is a SINT?
9. What is a BOOL? What is it typically used for?
10. What is an alias tag, and what is it used for?
11. What is the difference between a produced tag and a consumed tag?
12. What is an array tag? Write down an example of an array tag used to hold 7 speeds.
13. What is a structure-type tag?
14. What are the two types of tag scope? What are the differences between them?

15. What is the difference between an array tag and a structure tag?
16. What is a User-Defined tag? What is it used for?
17. More than one task can execute at one time. (True or False)
18. More than one program can execute at one time. (True or False)
19. What are the three execution types for tasks?
20. What is a watchdog timer?
21. How can the watchdog time be changed?

CHAPTER

2

Structured Text Programming

OBJECTIVES

Upon completion of this chapter, the reader will be able to:

- Explain the basics of the structured text language.
- List at least three benefits of structured text.
- Understand structured text routines.
- Utilize structured text to develop routines.

INTRODUCTION

Ladder logic has been the overwhelming choice for PLC programmers since PLCs were developed. Additional languages were specified by an international standard (IEC 61131-3). These languages are rapidly gaining in popularity and use. Structured text (ST) is one of the languages in IEC 61131-3. ST programming is more of a typical computer language compared with ladder logic. It is very similar to languages that are used in many industrial devices such as robots and vision systems. ST programming is also used within other PLC languages. The best way to learn the material in this chapter is to read the chapter and then work on the chapter questions before trying to write and test actual ST programs.

OVERVIEW OF STRUCTURED TEXT

ST language resembles C, Basic, Pascal, and even Visual Basic language. People are often most comfortable with the first programming language they learn or the one they have used the most. People who have used computer programming languages for other devices often believe that ST is the easiest language to use for programming PLC logic.

ST programs are written in short English-like sentences. This makes ST programs very readable and easy to follow and understand. ST programming is well suited for applications requiring complex mathematics or decision making. ST is also concise. An ST program for an application would be much shorter than a ladder logic program, and the ST logic would be easier to understand. Learning ST will help you program many industrial devices.

The benefits of ST are as follows:

- People who have programmed using a computer language can readily learn ST.
- Programs can be created in any text editor.
- ST runs as fast as ladder logic.
- It is concise and easy to understand.
- It can be used for all or a portion of an application's logic.

FUNDAMENTALS OF ST PROGRAMMING

Figure 2-1 shows a comparison between ladder logic and ST programming to accomplish a simple task.

Tags are used in the logic examples. The tags are named Temp, Flow, Pump, and Green_Light. These are all tags in a ControlLogix (CLX) project. This routine controls the pump, flow, and a green light.

If you study the ladder logic in Figure 2-1, you will see that it is a little confusing. It is not easy to understand what the logic is trying to accomplish. The logic in the ST is much easier to follow. It is more concise and direct in its approach. Note that you do not need to understand the logic completely at this point.

The ST routine uses IF statements to make decisions. The first portion of the logic compares the value of the Temp tag. If Temp is greater than or equal to 100 and less than 200, then Pump is turned on, the Flow variable (tag) is set to a value of 45, and Green_Light is turned on. Else if (ELSIF) Temp is less than 100, the pump is turned off, Flow is set to 20, and the Green_Light is turned off.

If neither of these conditions is true, then the Else is true and Alarm_Light is turned on.

ST logic is easier to understand than ladder logic. ST is not case sensitive. Tabs and carriage returns should be used to make programs clear and readable. They have no effect on the execution of the program, but they can really help make it more understandable. Indentation was used in the program in Figure 2-1 to make the program more understandable. It makes it much easier to follow the logic. Note that if this logic were created in a routine, it would have to be called from the main with a jump-to-subroutine (JSR) instruction for the routine to actually run.

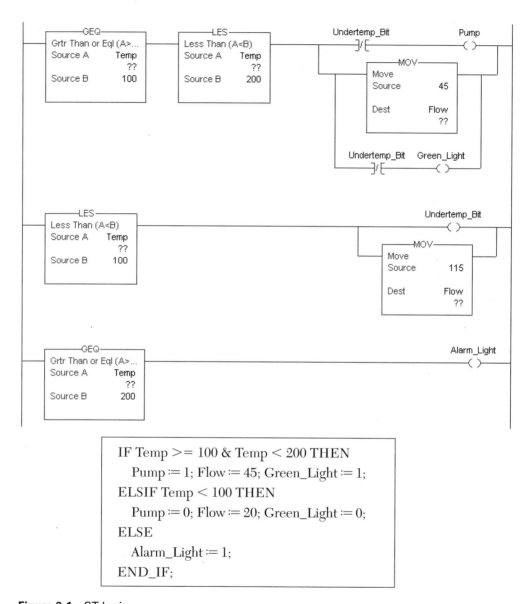

$$IF\ Temp >= 100\ \&\ Temp < 200\ THEN$$
$$\quad Pump := 1;\ Flow := 45;\ Green\_Light := 1;$$
$$ELSIF\ Temp < 100\ THEN$$
$$\quad Pump := 0;\ Flow := 20;\ Green\_Light := 0;$$
$$ELSE$$
$$\quad Alarm\_Light := 1;$$
$$END\_IF;$$

Figure 2-1   ST logic.

PROGRAMMING ST IN CONTROLLOGIX

An ST program is a routine within a CLX project. If you right click the mouse button on the main program icon, it will give you the option of adding a new routine. Figure 2-2 shows the initial screen for creating a new routine. In this example, the name Structured_ Text_Stop_Light was entered. Next the language Type is chosen from the drop-down menu in the Type box. In this example, Structured Text was chosen.

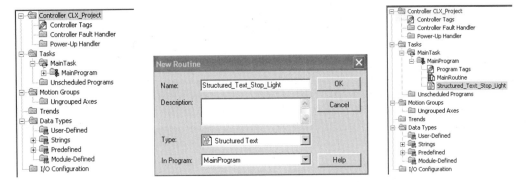

Figure 2-2 On the left is the Controller Organizer. If you right click on **MainProgram**, you can create a new routine. In the center is the New Routine configuration screen. Note that Structured Text was chosen for the language Type. The figure on the right shows the Controller Organizer after the new routine was created. After the new routine is chosen, you can click on the name of the routine in the main program list and the screen shown in Figure 2-3 will appear.

The program code can be entered into the ST entry screen. A very simple one-statement routine was entered in Figure 2-3. Green_1 is a tag that was created in the tag editor. Green_1 is a discrete output. This simple logic would set the value for Green_1 to 1 (true) when this routine is run. The output assigned to Green_1 output would be turned on. The := is used to change the value of whatever tag is on the left to the value shown on the right. This is called an assignment statement. Almost all program lines must end with a semicolon (;). This will be explained later in this chapter.

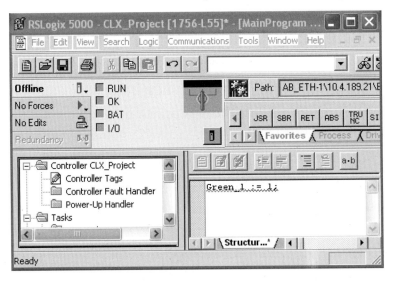

Figure 2-3 A simple program to turn output Green_1 on.

Note that to run, this routine would need to be called by the MainRoutine (see Figure 2-4). Note that a JSR instruction was used to run the ST routine.

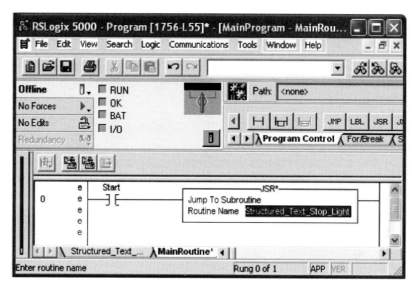

Figure 2-4 A JSR instruction being used to call (run) the ST routine. Note that in this example, the JSR will be executed only while the Start contact is true.

Note that a well-planned job is half done. It is wise to preplan the application and programming before you start writing the program. How many inputs and outputs will there be in the application? What tags will be required, and what types will they be? What would be appropriate names for all of the tags? It will ease your task if you create the tags before you begin to write a program. You are not required to create tags first, but it will make programming and troubleshooting less frustrating.

Assignment Statements

Assignment statements are used to assign a value to a tag. The example below would assign a value of 115 to the tag named Temp:

Temp := 115;

The generic example of an assignment statement is shown below.

Tag := value or mathematical expression;

The Tag on the left in the equation above represents a variable or tag that is being assigned a new value. A tag can be a Boolean (BOOL)-, a single-integer (SINT)-, an integer (INT)-, a double-integer (DINT)-, or a real (REAL)-type tag. The DINT is the default integer type and is the one that should normally be used for integers. The := is the assignment operator. Note that you cannot just use the equal sign. You must use an equal sign preceded by a colon. The expression on the right side of the equation is used to represent the value that will be assigned to the tag. An expression can be a constant value (a number), or it could be a tag of type BOOL, SINT, INT, DINT, or REAL. The last thing in the line is the semicolon. You must have a semicolon at the end of the line.

In fact, almost all lines in an ST program must be terminated with a semicolon. The most common error in ST programming is forgetting the semicolon.

Study the examples below. The first example shows a value of 212 assigned to the tag named Temp. The second example uses a tag (variable) to assign a value to Temp. The third example shows that math statements can also be used in assignment statements. In this example, Var_1 is being assigned the value of Var_2 multiplied by 2.

Temp := 212;

Temp := Var_1;

Var_1 := Var_2 * 2;

Documenting Logic with Comments

Comments should be used in programs. They help make programs more understandable. They also help reduce the time and frustration of troubleshooting an application. They help a technician understand what the programmer intended. A comment can appear anywhere in a program line or on a line by itself. Figure 2-5 shows the format for various types of comments. Figure 2-6 shows examples of some of the ways comments can be used.

To Use a Comment	Format
On a line without instructions Or At the end of a line of ST	// Comment (* Comment *) /* Comment */
Within a line of ST	(* Comment *) /* Comment */
On more than one line	(* Start of comment ....end of comment *) /* Start of comment ....end of comment */

Figure 2-5   Table showing possible format for comments.

Example	Examples of Different Ways to Add Comments in ST
1	// This is an example of a comment at the beginning of a line
2	IF Temp. 100 THEN // Comment at the end of a line
3	Temp := 212; /* Comment at the end of a line – different format */
4	Pump := 1; (* Comment at the end of a line – different format*)
5	IF S1 (* Comment within a line) & S2 (* Comment within a line) THEN
6	IF Temp2 = 205 /* Comment within a line – different format */ THEN
7	(* Comment on more than one line. This is an example of a comment that takes up more than one line in a program *)
8	/* Comment on more than one line. This is an example of a comment that takes up more than one line in a program in a different format */

Figure 2-6   Examples of the use of comments.

ARITHMETIC OPERATORS

All of the standard arithmetic operators are available in ST programming. Study the examples shown below. In the first line, the tag named Temp is assigned a value equal to Var1 plus 20. Line 2 shows a value assigned to the RPM tag that is equal to the Speed tag divided by 60. The third statement shows the result of the tag named Cases multiplied by 12 assigned to the Total_Cans tag.

Temp := Var1 + 20;

RPM := Speed/60;

Total_Cans := Cases * 12;

Figure 2-7 shows the arithmetic operators that can be used in ST programming. Math operators are most commonly used in assignment statements.

Instruction	Operator	Optimal Data Type
Add	+	DINT, REAL
Subtract	−	DINT, REAL
Multiply	*	DINT, REAL
Exponent (X to the power of Y)	**	DINT, REAL
Divide	/	DINT, REAL
Modulo	MOD	DINT, REAL

Figure 2-7 Arithmetic operators.

The programmer must be very careful to use correct number types when performing math operations. Integer math will not provide a decimal result. Study the example below. The tag Answer is a DINT type. Because it is an integer type, integer math will be done. You might expect the answer to be 2.5, but the answer would be 2. You would need to use a REAL type to get a decimal result. If the Answer tag is created as a REAL type, the answer would be 2.5.

Answer := 5/2;

Study the example below. You might expect the answer to be 2, but the answer would be 3. You must be careful when using integer (DINT type) math. If you want a decimal number as a result, use REAL-type tags.

Answer := 5.1/2;

Modulo Instruction

The modulo instruction is a very interesting operator. It can be used to find the remainder of a division. The result of a modulo operation is the integer remainder of the division.

Answer := 5 MOD 2; //Answer = 1   (5 MOD 2 = 2 with a remainder of 1)

Answer := 7 MOD 3; //Answer = 1   (7 MOD 3 = 2 with a remainder of 1)

Answer := 17 MOD 3; //Answer = 2   (17 MOD 3 = 5 with a remainder of 2)

Answer := 13 MOD 5; //Answer = 3   (13 MOD 5 = 2 with a remainder of 3)

ARITHMETIC FUNCTIONS

There are also many arithmetic functions available to the programmer. Figure 2-8 shows which functions are available. Note that the chart in Figure 2-8 also shows the optimal data type to use for each function. The example below shows the use of the square root function. In this example, the function would calculate the square root of 515 and assign the result to the tag named Val. Note that a tag (variable) could have been used in place of the constant 515.

Val := SQRT(515);

For	Function	Optimal Data Type
Absolute value	ABS(numeric expression)	DINT, REAL
Arc cosine	ACOS(numeric expression)	REAL
Arc sine	ASIN(numeric expression)	REAL
Arc tangent	ATAN(numeric expression)	REAL
Cosine	COS(numeric expression)	REAL
Radians to degrees	DEG(numeric expression)	DINT, REAL
Natural log	LN(numeric expression)	REAL
Log base 10	LOG(numeric expression)	REAL
Degrees to radians	RAD(numeric expression)	DINT, REAL
Sine	SIN(numeric expression)	REAL
Square root	SQRT(numeric expression)	DINT, REAL
Tangent	TAN(numeric expression)	REAL
Truncate	TRUNC(numeric expression)	DINT, REAL

Figure 2-8   Arithmetic functions.

RELATIONAL OPERATORS

Relational operators are used to compare two values or strings and provide a true or false result. The result of a relational operation is a BOOL value. If the result of an operation is true, the result will be a 1. If the result of a relational operation is false, the result will be 0. These are used extensively for decision making. Figure 2-9 shows a table of relational operators.

Relational operators can be used to make decisions. For example, an IF statement can make use of relational operators.

IF TEMP > 200 THEN

In this example, the value of TEMP is evaluated to see if it is greater than 200; if it is, this evaluates to a value of 1 (true). If TEMP is less than 200, it would have a value of 0 and be false. The THEN would be executed only if the value of the operation is a 1 (true).

Comparison Type	Operator	Optimal Data Type
Equal	=	DINT, REAL, string
Less than	<	DINT, REAL, string
Less than or equal	<=	DINT, REAL, string
Greater than	>	DINT, REAL, string
Greater than or equal	>=	DINT, REAL, string
Not equal	<>	DINT, REAL, string

Figure 2-9   Relational operators.

The value of a relational operation can also be assigned to a tag. In the statement below, if TEMP is greater than or equal to 200, the value of 1 will be assigned to the tag called STAT. If TEMP is less than 200, the value of 0 will be assigned to the tag named STAT.

STAT := (TEMP >= 200);

Relational operators can also be used to evaluate strings of characters or characters in strings. In the example below, the = relational operator is used to evaluate whether String_1 is equal to the second string (Password). If they are the same, the result will be a 1 (true). If the strings are not equal, the result will evaluate to a 0 (false).

IF String_1 = Password THEN

The example below would evaluate the first character in a string called String_1 to see if it is equal to 65. The letter A in ASCII is equal to 65. The String_1.DATA[0] represents an array of characters. The 0 in square brackets represents the character we want to evaluate.

IF String_1.DATA[0] = 65 THEN

LOGICAL OPERATORS

Logical operators can also be used to check to see if multiple conditions are true. Figure 2-10 shows a table of logical operators.

Type	Operator	Type
Logical AND	&, AND	BOOL
Logical OR	OR	BOOL
Logical exclusive OR	XOR	BOOL
Logical complement	NOT	BOOL

Figure 2-10   Logical operators.

In the example below, Sensor_1 is evaluated. If the sensor is on, it would evaluate to a 1 (true) and the THEN would be executed.

IF Sensor_1 THEN

In the example below, Sensor_1 is evaluated. In this case, a NOT was used. So, if the Sensor_1 is a 0 (false), the THEN would be executed.

IF NOT Sensor_1 THEN

In the next example, both must evaluate to true for the whole statement to be evaluated as true. Note that the & operator was used for the AND logical operator. AND or & can be used.

IF Sensor_1 & (TEMP < 150) THEN

In the next example, the OR logical operator is used. In this case, if either Sensor_1 or Sensor_2 is true, the statement will evaluate to a 1 and the THEN will be executed.

IF Sensor_1 OR Sensor_2 THEN

An Exclusive-OR (XOR) is used in the next example. In this example, only one of the two sensors can be true for the statement to be evaluated to true.

IF Sensor_1 XOR Sensor_2 THEN

In the next example, the result (1 or 0) of the logical operation will be assigned to the tag called STATUS. If Sensor_1 and Sensor_2 are both true, the tag STATUS will be assigned a value of 1.

STATUS := Sensor_1 & Sensor_2

PRECEDENCE

The table in Figure 2-11 shows the order in which math statements will be evaluated. This is very important. The wrong answer will be obtained if precedence is not carefully considered. Consider the example below. Normally math statements are evaluated from left to right if precedence is equal. In the example below, precedence is not equal. Many people would say the answer is 21. They might add the 5 and the 2 (7) and then multiply by 3, getting a result of 21. The correct answer is 11. Multiplication has a higher precedence than addition. You must multiply 2 * 3 (6) and then add the result to 5 (11).

Answer := 5 + 2 * 3;

Another method to assure the proper order of calculation is to use parentheses.

In the example below, it was desired to do addition first. It will not be done first because the multiplication operator has a higher precedence than the addition operator.

Answer := Var1 + 17 * Temp;

It could be rewritten as shown below. In this example, the parentheses assure that addition will be done first. Parentheses have the highest precedence.

Answer := (Var1 + 17) * Temp;

Order	Operation
1	( )
2	Function ( )
3	**
4	− (negate)
5	NOT
6	*, /, MOD
7	+, − (subtract)
8	<, <=, >, >=
9	=, <>
10	&, AND
11	XOR
12	OR

Figure 2-11   The order of precedence for arithmetic operators.

CONSTRUCTS

The definition of *construct* is to form by assembling or combining parts, to build. The table in Figure 2-12 shows the constructs that are available in ST. A construct can also be thought of as a statement.

Construct	When to Use
IF THEN	If specific conditions are true
FOR DO	A specific number of times
WHILE DO	As long as a condition is true
REPEAT UNTIL	Until a condition is true
CASE OF	On the basis of a number

Figure 2-12   Constructs that can be used in ST.

An IF statement can be used to make a decision and then execute logic on the basis of the decision. The IF is followed by a test statement. The test statement is a Boolean expression that is evaluated to be true (1) or false (0). In the example below, if the BOOL expression is true, all statements between the THEN and the END_IF will be executed. If the BOOL expression is false, processing would continue after the END_IF.

```
IF BOOL_expression THEN
    Motor_1 := 1;
    Temp := 150;
    Additional logic
END_IF
```

An ELSE may also be used. In the example shown below, if the tag Motor_On is true, the tag Red_Light will be set to 1. If Motor_On is false, the tag Green_Light will be set to 1. Note that semicolons are very important. There is no semicolon after the THEN or the ELSE, but every other line is terminated with a semicolon.

```
If Motor_On THEN
    Red_Light := 1;
ELSE
    Green_Light := 1;
END_IF;
```

ELSE IF (ELSIF) Statements

The example below shows the use of an ELSIF statement. Note the spelling of the ELSIF. In this example, if the first IF is false, the ELSIF is evaluated. If Sensor_3 is true, Alarm will be assigned the value 1. Note also that you must have an END_IF statement.

```
IF Sensor_1 & Sensor_2 THEN
    Pump := 1;
    Heat_Coil := 0;
ELSIF Sensor_3 THEN
    Alarm := 1;
END_IF;
```

The example below adds a few new twists. An IF, an ELSIF, and an ELSE are used. In this example, if the IF is false and the ELSIF is false, the ELSE will be executed and the Alarm_Light tag will be assigned the value 1. Note that in this example more than one statement is put on a line. Each is separated by a semicolon. While this is permissible, it should not be used to excess. You should be careful that such use does not make your program more difficult to understand. Note also that the statements are indented to make the program easier to understand. Finally, note where semicolons are used and where they are not used.

```
IF Temp >= 100 & Temp < 200 THEN
    Pump := 1; Flow := 45; Green_Light := 1; Alarm_Light := 0;
ELSIF Temp < 100 & Temp > 50 THEN
    Pump := 0; Flow := 20; Green_Light := 0; Alarm_Light := 0;
ELSE
    Alarm_Light := 1; Pump := 0; Flow := 0; Green_Light := 0;
END_IF;
```

FOR DO Statements

A FOR DO loop is used when we know how many times a loop should be executed. For example, if we needed to fill an array of 10 integers with a number, we could use a FOR DO loop that would execute 10 times.

A FOR loop uses a variable to increment each time through the loop. In the example below, a variable named X is first set to a value of 0 by the loop. X is incremented each time the loop executes. So the second time the loop executes, X is incremented by 1; the third time it executes, X is incremented by 2; and so on. The "by 1" in the loop makes the loop increment by 1 each time it executes. When X increments by 10, the loop will not execute any more.

In the FOR loop example below, a tag named Temp was created and it was configured to be an array of 10 tags named Temp (Temp[0] through Temp[9]). The FOR loop below is used to fill the array of 10 tags (Temp[0] through Temp [9]) with the number 99.

```
FOR X := 0 to 9 by 1 DO
    Temp[X] := 99;
END_FOR;
```

In the next example, the loop was incremented by 6 (by 6) each time through. X in the loop is assigned a value of 0 the first time through the loop. The next time through the loop, it is assigned a value of 6, then 12, then 18, then 24. The next time it would be 30, but 30 is greater than 24, so the loop would not execute again.

```
FOR X := 0 to 24 by 6 DO
    Additional ST Statements;
    Additional ST Statements;
END_FOR;
```

Note that when a loop is executed, the controller does not execute any other statement in the routine outside of the loop until the loop is completed. If the time it takes to complete a loop is greater than the watchdog timer for the task (default time is 500 ms), a major fault will occur. If this becomes a problem, try a different type of construct, such as an IF THEN.

WHILE DO Statement

A WHILE DO construct is used when you would like something to happen *while* a condition is true. In the following example, we would like the loop to execute WHILE two things are true: the tag CNT is less than 50 AND the tag TEMP is less than 200. If CNT becomes greater than 49 or Temp becomes greater than 199, the loop will not run. Note that integer numbers are used.

```
WHILE ((CNT < 50) & (Temp < 200)) DO
    CNT := CNT + 3;
END_WHILE;
```

Remember that when a loop is executed, the controller does not execute any other statement in the routine (outside of the loop) until the loop is completed. If the time it takes to complete a loop is greater than the watchdog timer for the task, a major fault will occur. If this becomes a problem, try a different type of construct, such as an IF THEN. Note that with the WHILE loop, the test is at the beginning of the loop. If the test is false, the loop will not even execute once.

REPEAT UNTIL Statement

The REPEAT UNTIL statement is used when we have something we want to do UNTIL certain conditions are true. In this example, the REPEAT will execute UNTIL one of two conditions is true (note the use of the OR). In this example, 2 is added to the value of the tag CNT every time the REPEAT is executed. The REPEAT will end when CNT is greater than 34 OR Temp is greater than 92. Note where semicolons are used and where they are not used.

```
REPEAT
    CNT := CNT + 2;
    UNTIL ((CNT > 34) OR (Temp > 92))
END_REPEAT;
```

Note that a REPEAT loop will execute at least once, because the test is at the end of the loop. This can cause confusion. Remember that the logic will run at least once every time it is called by a JSR because the test is at the end of the loop.

CASE OF Statement

The CASE OF construct is very useful. It can be used to perform different portions of the program depending on a value. For example, we might produce six different products on one machine. We could have the operator enter a number between 1 and 6 for the product that needs to be made. The CASE statement would then run the section of code to produce that product.

The value of the CASE variable (tag) determines which section of the program is executed. In the example below, the value of tag Var_1 is used to decide which part of the logic to execute. This example shows several ways in which more than one number can be used to select a section of code.

> If Var_1 is equal to 1, Temp_1 will be assigned a value of 85 and Pump_1 will be turned on.
>
> If the value of Var_1 is 2, Temp_1 will be assigned a value of 105 and Pump_1 will be turned on.
>
> If Var_1 is equal to 3 or 4, Temp_1 will be assigned a value of 110 and Pump_1 will be turned on.

Note the use of a comma. When a comma is used between values, each is valid. For example, 1, 4, 7, or 9 could be used. In that case if Var_1 were equal to a 1, 4, 7, or 9, that portion of the program would be executed.

If the value of Var_1 is a 5, 6, 7, or 8, Temp_1 will be assigned a value of 115 and Pump_1 will be turned on. Note the use of the periods between the 5 and the 8. This means that any numbers between the two specified numbers are also valid. In this case, a 5, 6, 7, or 8 would cause that portion of the program to be executed.

In the next portion of the code, both the comma and the period declarations have been used. In this case, if Var_1 is a 9, 11, 12, 13, 14, or 15, this portion of code will be executed.

Note the use of the ELSE at the end. If Var_1 has a value that is different from any of the cases specified, the ELSE portion of code will be run.

```
CASE VAR_1 OF
1:          Temp_1 := 85;
            Pump_1 := 1;
2:          Temp_1 := 105;
            Pump_1 := 1;
3,4:        Temp_1 := 110;
            Pump_1 := 1;
5...8:      Temp_1 := 115;
            Pump_1 := 1;
9,11...15:  Temp_1 := 120;
            Pump_1 := 1;
ELSE
    Alarm := 1;
    Additional logic;
END_CASE;
```

TIMERS

Other traditional programming instructions can be used in ST. For example, some types of timers, such as a retentive (TONR) time, can be used. To check which instructions are available and to see an example of their use, you can use the instruction help available in

RSLogix 5000. Look up a particular instruction, and it will explain which languages it can be used in and will also give examples of the use of the instruction in the language. When a timer tag is created, several tag members are created automatically. For example. If a TONR timer is created and given the tag name of Stop_light_Timer, several tag members will also be created. Note that in ST the tag type must be an *FBD Timer*. Figure 2-13 shows the tag and tag members for the timer.

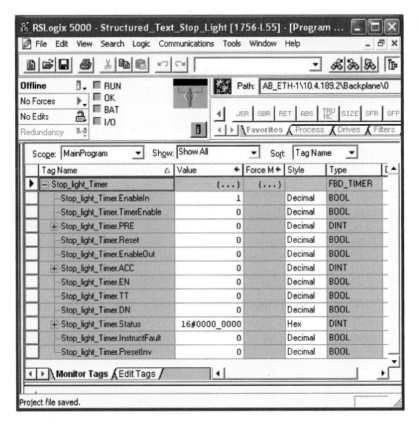

Figure 2-13   Timer tag members created when you create a timer.

The next example shows how a timer can be used in ST programming. Study the example. Note the use of the preset (.PRE), accumulated time (.ACC), and the timer enable (.TimerEnable). The first three lines are used to call the timer, set a PRE value, and enable the timer (start the timer timing). The next portion of the code is IF statements that turn a green light on if the timer's ACC value is greater than 0 and less than 10,000. The last IF statement resets the timer if the accumulated time has reached 30,000 ms by making the timer enable 0 (false).

```
TONR(Stop_light_Timer);
Stop_light_Timer.PRE := 30,000;
```

Stop_light_Timer.TimerEnable := 1;
IF (Stop_light_Timer.ACC > 0 & Stop_light_Timer.ACC < 10,000) THEN
   Green_1 := 1;
ELSE
   Green_1 := 0;
END_IF;
IF Stop_light_Timer.ACC = 30,000 THEN
   Stop_light_Timer.TimerEnable := 0;
END_IF;

Figure 2-14 shows what this program would look like in RSLogix 5000. Note also the Watch list at the bottom of the screen. This enables the programmer to watch the values of tags during run mode.

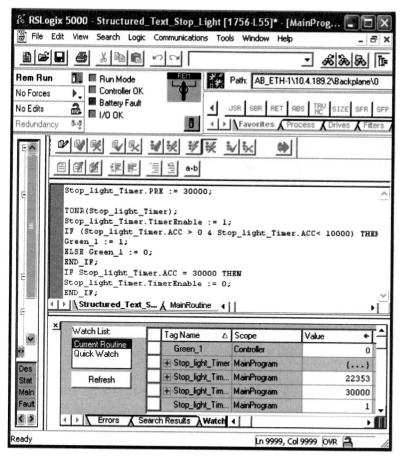

Figure 2-14 ST program example.

QUESTIONS

1. Write an assignment statement to:

   Assign the value of 5 to a tag named Count.

   Assign the value of the tag named T5 to a tag named Value.

   Multiply the value of a tag named count by a tag named Number and assign the result to a tag named Count.

2. Write one comment for each of the following. Make sure you use the correct format for a comment and make sure you explain the line. Use more than one format type.

   Temp := An_In_1/994.3;

   Var1 := 7;

   Total := ((5 + 8)/7) + 2;

3. Write ST for the following:

   Add 20 to a tag named TEMP and assign the result to Curr_Temp.

   Assign the value of a tag named Total to a tag named Amount.

   Multiply two tags and assign the result to a new tag.

   Motor_1 is a tag for digital output. Turn it on with an assignment statement.

4. What is the difference between DINT-type arithmetic and REAL-type arithmetic?

5. What is the result of each of the following statements?

   Answer := 4/2 /* Answer tag is a DINT */

   Answer := 5/2 /* Answer tag is a DINT */

   Answer := 5/2 /* Answer tag is a REAL */

   Answer := 5 MOD 2 /* Answer tag is a DINT */

6. Write a line of ST for each of the following:

   Find the square root of a number and assign the value to a tag.

   Find the tangent of a tag and assign it to another tag.

   Square a tag and assign the result to another tag.

7. Write an IF statement for each of the following:

   Temp is greater than 250.

   Var1 is greater than or equal to Var2.

   Var2 is less than Var1 multiplied by 6.3.

   If Sensor_1 is on, turn on Light_1.

   If Temp > 95, turn heater_1 off.

   If Temp is greater than 100 and Sensor_1 is true, turn Done_Light on; otherwise turn Alarm on.

8. When should a FOR DO loop be used?

9. Write a FOR DO loop to fill an array of temperatures with zeros.

10. Write a WHILE DO loop to turn Alarm_1 on until Temp is less than 150.

11. When should a REPEAT UNTIL loop be used?

12. Thoroughly explain the following ST:

```
CASE Choice OF
1:          V ar1 := 65;
            Out_1 := 1;
2,3:        Var1 := 85;
            Out_2 := 1;
4:          Var1 := 95;
            Out_3 := 1;
5,7 ... 10: Var1 := 115;
            Out_4 := 1;
ELSE
    Statement;
    Statement;
END_CASE;
```

13. Write an ST to use a 60-s timer. An output should be on for the first half of the cycle and off for the second half of the cycle. It should continuously repeat.

14. Where can you find which CLX instructions are available for ST programming?

15. Write the ST code for the following application:

Develop a stoplight application. You must program both sets of lights. Make the overall cycle time 30 s.

Tag name	Description
Green_1	Green—East/West
Yellow 1	Yellow—East/West
Red_1	Red—East/West
Green_2	Green—North/South
Yellow_2	Yellow—North/South
Red_2	Red—North/South

CHAPTER

3

Sequential Function Chart Programming

OBJECTIVES

Upon completion of this chapter, the reader will be able to:

- Explain what sequential function chart programming is.
- Explain types of applications that could benefit by the use of sequential function chart programming.
- Develop sequential function chart programs.

INTRODUCTION

Sequential function chart (SFC) programming is a very useful and user-friendly language. SFC is very useful in helping organize an application. This chapter will start with an overview of what an SFC program is and then explain each of the components of an SFC program in detail. It should be noted that the only way you will learn to program is by writing programs. Utilize the questions at the end of the chapter to make sure you understand the concepts and then practice writing and testing your logic.

SFC PROGRAMMING

SFC programming is a graphically oriented programming language. An SFC program looks like a flow diagram or a decision tree. If an application is sequential in nature, SFC programming is a natural choice. In SFC programming, an application is broken into logical steps. For example, when making a pot of coffee, there are some definite steps that are followed.

1. Measure and add coffee.
2. Measure and add water.

3. Turn the pot on.
4. Remove completed coffee and turn the pot off.
5. Remove the filter and grounds and clean the pot.

SFC programming could be used to program this system. The coffee-making process could be broken into five steps. Figure 3-1 shows an example of what it might look like. Step 1 has an *action* that the operator must perform. The action is to measure and add coffee. There is a decision at the end of the step. If coffee has been added, the process continues at step 2; if not, processing remains in step 1. Step 2 has one action associated with it (measure and add water). At the end of step 2 is another decision. Step 3 also has an action and a decision. Steps 4 and 5 have two actions associated with each. Note that every step has a decision point after it that determines when processing for the step is done and the next step should be started.

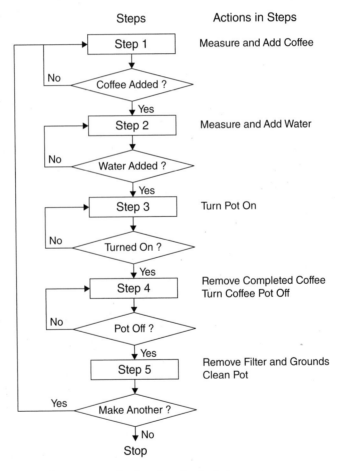

Figure 3-1 Coffee-making process broken into logical steps.

An SFC program consists of three main components: steps, actions, and transitions. Carefully study the typical SFC program in Figure 3-2. In this example, the first two step names were left with their default name: Step_000 and Step_001. The remaining three steps were given names that reflect the purpose of the steps: Normalize, Assemble, and Paint. There are transitions between steps (Tran_000, Tran_001, and Tran_002).

Study the transition between Step_000 and Step_001. A transition is a condition to determine when processing moves from one step to the next by evaluating to true or false. If the transition is true, processing will move to the next step. The first two steps are linear. In other words, the first one is processed, then the second, and then the next. The next three steps are concurrent steps. All three are being processed at once. Note the actions that are attached to steps (Action_000, Action_001, and Action_002). Actions can use structured text (ST) to make decisions and control I/O. Note that actions can be hidden to make a program less cluttered. Note also that each action has a qualifier that controls when an action starts and stops. These will be covered later in this chapter.

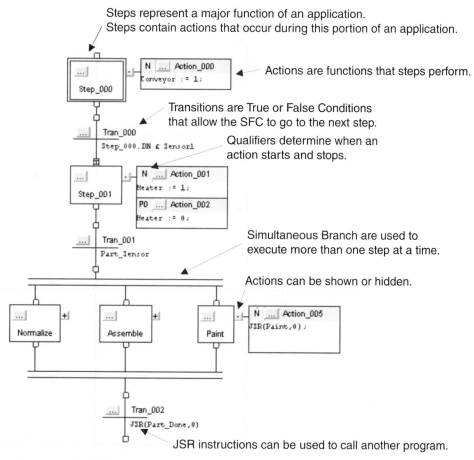

Figure 3-2 Simple SFC program.

SAMPLE APPLICATION

For this application, imagine a system that is used to heat parts to a specified temperature.

- The operator puts a part in the fixture and pushes a start switch.
- The heating coil is turned on for 50 s.
- When 50 s is done, the operator removes the completed part, places another part in the fixture, and starts a new cycle.

Figure 3-3 shows two examples of simple SFC programs that could control the heating application. Step 1 is named Wait_For_Start_Cycle. The processor will stay in this step until the transition after the step is true. In this example, the transition condition is the state of the start switch (Start_Switch). When the operator pushes the start switch, the transition is true and the processor will execute the next step (Heat_Cycle).

Heat_Cycle is a timed step. Parameters were set in the step to make it 50-s steps. An action was added to the step. The action turns on the heater output (Heat_Output). The heater output was programmed to be nonretentive, so it will shut off when the processor leaves this step. When the step has reached 50 s, the Heat_Cycle.DN bit will be true. Note that this bit was used for the next transition. In the example on the left, the transition was then wired to the first step, so the process can be executed again when the start switch is pushed by the operator. The example on the right is almost the same except it will run only once. A stop was programmed at the end of the sequence.

Steps are the logical groupings that we break our application into; for example, the first step in our coffee-making process is measuring and adding coffee. The step is the

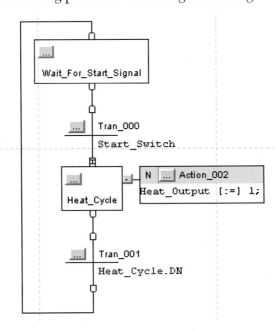

Figure 3-3 Simple heating cycle SFC programs. The one on the top will repeat. The one on the bottom will run once and stop.

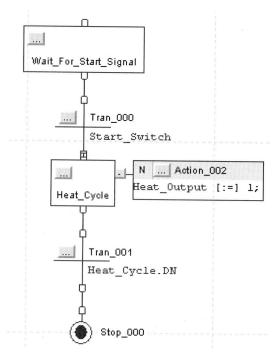

Figure 3-3 *Continued*

organizational unit. Steps can perform timing and counting functions and can have actions associated with them.

Actions can be thought of as the inputs and outputs we use to accomplish the tasks in a step. Actions in a step are repeated until the transition to the next step becomes true. Transitions are BOOL statements that must be true to move to the next step.

ORGANIZING THE EXECUTION OF THE STEPS

Linear Sequence

A linear sequence is used to execute one or more steps in a linear fashion. Figure 3-4 shows a bottling operation that is linear.

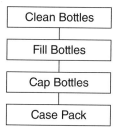

Figure 3-4 A linear step bottling process.

Figure 3-5 shows a linear SFC. One step is executed continuously until the transition becomes true; then the next step in the sequence is executed continuously until the next transition is true, and so on.

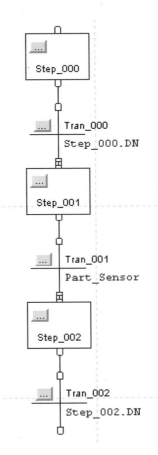

Figure 3-5   Linear steps in an SFC.

Wiring (Connecting) Steps

Wires are used to connect steps. Wiring is done by dropping elements on the attachment tabs or by clicking on the tab of one element and dragging it to the tab of the element you want to connect to. Tabs turn green when you can connect. You can connect a step to a previous point in an SFC (see Figure 3-6). This enables you to loop back to repeat steps or return to the beginning of an SFC and start over.

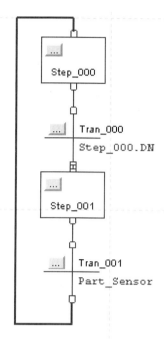

Figure 3-6   Looping back in an SFC.

Concurrent (Simultaneous) Processing

If there were more than one person making coffee, they could do some steps at the same time. One person could be putting a filter and coffee into the coffee machine, while another could be measuring water and pouring it into the pot. Figure 3-7 shows an example of concurrent coffee processing.

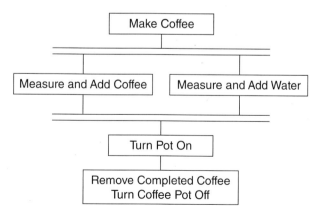

Figure 3-7   Concurrent processing.

A simultaneous branch is used to execute two or more steps or groups of steps at the same time. Figure 3-8 shows an example of a simultaneous branch. All paths must finish before continuing on in the SFC program. A single transition is used to enter a simultaneous branch and a single transition is used to end a branch. The SFC program will check the end transition, after the last step in each path has executed at least once. If the transition is false, the last step is repeated.

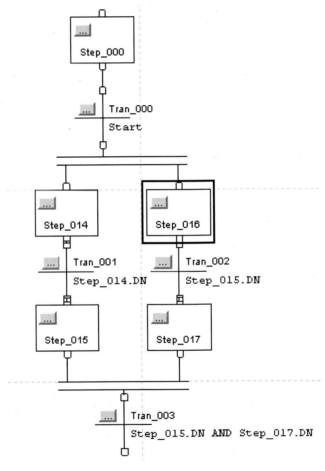

Figure 3-8 A simultaneous branch. Note the parallel horizontal lines at the top and bottom of the branch.

Selection Branching

A selection branch is used to choose between paths of steps depending on logic conditions. Figure 3-9 shows a selection branch SFC. Note that each path begins with a transition. The SFC program will check the transitions that start each path from left to right. It will take the first path that is true. If no transitions are true, the previous step will be repeated. The software will let you change the order in which transitions are checked. It is acceptable for a path to have no steps and only a transition. This could be used to skip an entire selection branch under certain conditions. In Figure 3-9, if the first three transitions are false and Tran_015 is true, processing would go through the selection branch to the Pack step after the branch.

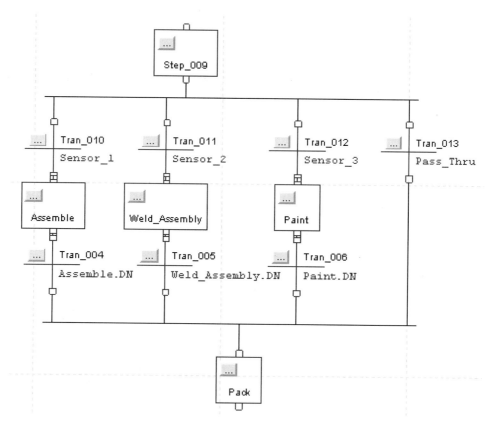

Figure 3-9   A selection branch SFC. Note the single horizontal lines at the beginning and end of the selection branch.

At this point, you should understand the basics of what an SFC is. You should understand what steps, actions, and transitions are used for, which in the rest of the chapter steps, actions, and transitions will be examined in more detail.

STEPS

Steps have a lot of functionality built into them. To identify steps in a process, it may be helpful to look for physical changes in the process. A physical change might be a pressure that is reached, a temperature that is reached, a new part that is now in position, or a choice of which product or recipe to use. This physical change can represent the end of a step. The step will consist of the actions that occur before that change. Be careful not to have too many steps or too few steps. Too many or too few may make the program confusing. Make steps meaningful.

When a step is created, several tag members are automatically created that are associated with the step's tag name. Figure 3-10 shows an example of members that were automatically created for a step tag named Step_000.

⊞ Step_000.Status
Step_000.X
Step_000.FS
Step_000.SA
Step_000.LS
Step_000.DN
Step_000.OV
Step_000.AlarmEn
Step_000.AlarmLow
Step_000.AlarmHigh
Step_000.Reset
⊞ Step_000.PRE
⊞ Step_000.T
⊞ Step_000.TMax
⊞ Step_000.Count
⊞ Step_000.LimitLow
⊞ Step_000.LimitHigh

Figure 3-10 Step_000 tag members.

Let's consider a few of the available tag members.

> Step_000.DN would be the done bit. If we set a PRE value in milliseconds for the step, the DN bit for the step will be true when the step's time gets to the PRE value that is found in Step_000.T.
>
> Step_000.AlarmHigh is a bit that would be set to true when the step's accumulated time (Step_000.T) gets to Step_000.LimitHigh value. Any of the tag members can be used in logic.

You can change the name of the step by right clicking on the step and choosing the Rename option. When you rename a step, all of the step tag members are renamed to the new name. One of the tags is a DN bit.

Figure 3-11 shows a table listing all of a step's members and their use. These step members (tags) can all be used in logic. To use tag members, the name of the step is followed by a period (.) and the member name.

To	Tag Member	Data Type	Function
Length of time a step has been active	T	DINT	When the step becomes active, the T value is reset and then begins to increment in milliseconds.
Length of the timer preset	PRE	DINT	Enter the length of time you want for the timer PRE value.
Timer DN Bit	DN	BOOL	The DN bit is set when the timer reaches the PRE value. The DN bit will stay on until the step becomes active again.
If a step did not execute long enough	LimitLow	DINT	Enter a value in milliseconds. If the step becomes inactive before the timer (T) reaches the LimitLow value, the AlarmLow bit turns on. The AlarmLow will stay on until you reset it. To use the alarm function, make sure the alarm enable (AlarmEn) is checked.
	AlarmEn	BOOL	To use the alarm bits, you must turn (check) on the alarm enable (AlarmEn) bit in the step.
	AlarmLow	BOOL	If the step becomes inactive before the timer (T) has reached the LimitLow value, the AlarmLow bit will be turned on. The bit will stay on until you reset it. To use this alarm function, you must turn on (check) the alarm enable (AlarmEn) bit in the step.
If a step is executing too long	LimitHigh	DINT	Enter a value in milliseconds. If the step becomes inactive before the timer (T) reaches the LimitHigh value, the AlarmHigh bit turns on. The AlarmHigh will stay on until you reset it. To use the alarm function, make sure the alarm enable (AlarmEn) is checked.
	AlarmEn	BOOL	To use the alarm bits, you must turn (check) on the alarm enable (AlarmEn) bit in the step.
	AlarmHigh	BOOL	If the step becomes inactive before the timer (T) has reached the LimitHigh value, the AlarmHigh bit will be turned on. The bit will stay on until you reset it. To use this alarm function, you must turn on (check) the alarm enable (AlarmEn) bit in the step.
Do something while the step is active (including the first and last scan)	X		The X bit is set the entire time a step is executing. It is recommended to use an action with an N (nonstored) qualifier to do this.

Figure 3-11 Tag members for a step.

Do something once when the scan becomes active	FS	BOOL	The FS bit is on during the first scan of a step. It is recommended to use an action with a P1 Pulse (rising edge) to do this.
Do something while the step is active	SA	BOOL	The SA bit is on while a step is executing except for the first and last scan.
Do something one time on the last scan of the step	LS	BOOL	The last scan (LS) bit is on during the last scan of a step. This bit should only be used if you do the following: In the controller's properties box, SFC execution tab, set the Last Scan of Active Step to Don't Scan or Programmatic Reset.
Determine the target of an SFC Reset (SFR) instruction	Reset	BOOL	An SFR instruction resets the SFC to a step or stop that the instruction specifies. The Reset bit indicates to which step or stop the SFC will go to begin executing again. Once the SFC executes, the Reset bit is cleared.
The maximum time that a step was active during any execution	TMax	DINT	This is normally used for diagnostics. The controller will clear this value only when you select the Restart Position of Restart at initial step and the controller changes modes or experiences a power cycle.
Determine if the timer (T) value rolls over to a negative value	OV	BOOL	This is used for diagnostics.
Find out how many times a step has been active	Count	DINT	This is the number of times a step has been active, not the number of scans. The count is incremented each time a step becomes active. It will only be incremented after a step goes inactive and then active again. The count only resets if you configure the SFC to restart at the initial step. Under this configuration it will be reset when the controller is changed from program to run mode.

Figure 3-11 *Continued*

Using the Preset Time of a Step

The PRE value of a step can be used to control how long a step executes. Figure 3-12 shows the properties screen for a timed step. In this example, the heating step (Step_000) needs to be run for 40 s (40,000 ms). Note that the PRE value for a step is entered into the Preset member. Figure 3-13 shows the step and transition. The DN bit of the step is used as the transition to quit executing Step_000 and move to the next step. Note that the Step_000 DN bit was used (Step_000.DN) in the transition. When the step's accumulated time equals 40,000, the Step_000.DN bit will be true, making the transition true, and the program will move to the next step.

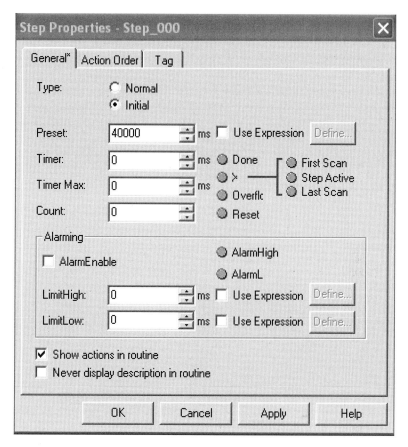

Figure 3-12   Step properties.

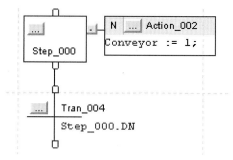

Figure 3-13   A timed step.

Use of a Step's Alarm Members

The next example makes use of the Step's AlarmHigh member. The AlarmHigh member is one bit. Study Figure 3-14. To use the alarm bits in a step, you must turn on the Alarm Enable (AlarmEn) parameter bit in the step (refer back to Figure 3-12). This is done in the check box labeled AlarmEnable. You must also enter a value for LimitHigh. If the step's timer (T) has reached the LimitHigh value, the AlarmHigh bit will be turned on. The bit will stay on until you reset it. In this example, the step was supposed to be finished successfully in less than 10 s (10,000 ms). If it is finished, the transition on the left of the selection branch will be true and the step named Process will be executed. If the tasks in the step are not completed in less than 10 s, the step's AlarmHigh bit (INIT. AlarmHigh) will be set, the transition on the right will be true, and the step named Shutdown will be executed.

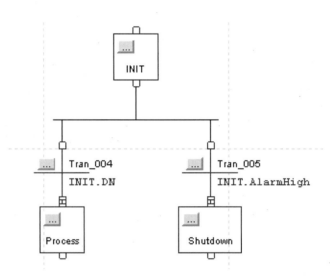

Figure 3-14   Use of a Step's AlarmHigh member (bit).

Turning Devices Off at the End of a Step

Devices can be turned off at the end of a step through program logic or automatically. There are three choices for how steps are scanned. The choices are found in Controller Properties under the SFC Execution tab (see Figure 3-15). The three choices that control how the last scan of an active step is handled are Automatic reset, Programmatic reset, and Don't scan.

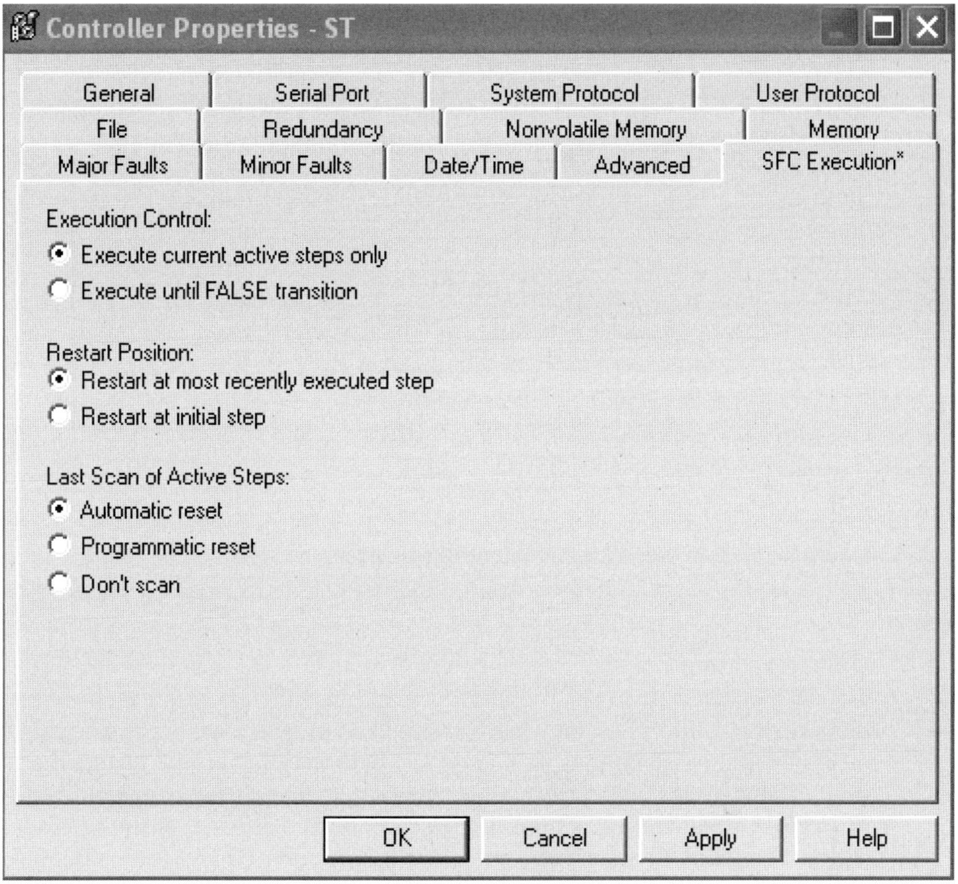

Figure 3-15   Controller Properties, SFC execution tab.

Automatic Reset Option
Automatic reset may be the most straightforward of the three. If you check the Automatic reset option choice (see Figure 3-15) and square brackets in assignment statements as shown in Figure 3-16, outputs will be turned off when leaving the step. If the square brackets were not used, the output would remain on.

Output _5 [:=] 1;

The square brackets would make this action nonretentive, and Output_5 would be turned off when the step ends. Remember that you must choose Automatic reset in the Controller Properties for this to work.

Programmatic Reset Option
If the Programmatic reset option is chosen, you can use the last scan of a step to change the state of devices. Figure 3-16 shows an example of Programmatic reset. Note that the step's LS bit is used in the IF statement. During all scans except the last scan, the output is on. On the last scan of the step, the output would be turned off.

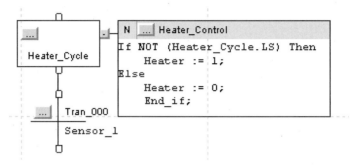

Figure 3-16 Use of the LS bit to control a device.

Don't Scan Option
The Don't scan is the default option for scanning. If this option is chosen, all data keep their current value when they leave a step. The programmer must use assignment statements or other instructions to change any data that need to be changed at the end of a step. A falling edge pulse (P0) action can be used. The P0 action should be the last action in a step. Only P and P0 actions are executed in the last scan if the Don't scan option is chosen. Figure 3-17 shows an example of the use of a P0 in Action_002 to turn a device off at the end of a step.

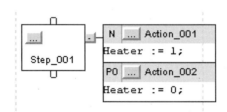

Figure 3-17 Use of a P0 action to turn an output off at the end of a step.

These have been only a few examples of what can be done with the step tag members. Study the table of step tags and their uses in Figure 3-11 for additional possibilities.

ACTIONS

Actions are used to perform functions such as turning outputs on or off in a step. Actions are added to steps by right clicking on a step and then choosing Add Action. Two actions have been added to a step in Figure 3-18.

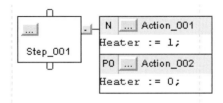

Figure 3-18   Actions in a step.

Action Tag Structure

Action tag members are created when you create an action. Figure 3-19 is a table that shows the action tag members and their uses.

To	Check or Set This Member of the Structure	Type	Details
Determine when the action is active.	Q	BOOL	The status of the Q bit depends on whether the action is a Boolean or a non-Boolean action.
			Type of Action / State of Q Bit
			Non-Boolean — On while the action is active, but off at the last scan of the action
			Boolean — On the entire time the action is active including the last scan of an action
			Use the Q bit to determine when an action is active.
	A	BOOL	The A bit is true the entire time an action is active.
Determine how long an action has been active in milliseconds.	T	DINT	When an action becomes active, the timer (T) value resets and then starts to count up in milliseconds. The timer will count up until the action goes inactive, regardless of the PRE value.
Use a time-based qualifier such as L, SL, D, DS, or SD.	PRE	DINT	Enter a time limit, or delay, in the preset (PRE) member. The action starts or stops when the timer (T) reaches the PRE value.
Determine how many times an action has become active.	Count	DINT	Count is not a count of the scans of the action. The count is incremented each time the action becomes active. Count will increment only when the action goes inactive and then active again. The count will only reset if the SFC program is configured to restart at the initial step. With that configuration, it resets when the controller changes from program to run mode.

Figure 3-19   Action tag member table.

There are two types of actions, non-Boolean and Boolean.

Non-Boolean Actions

Non-Boolean actions contain the logic for an action. They use ST to execute instructions or call a subroutine. ST can be used in actions for assignment statements, logic, or instructions. Figure 3-20 shows an example of an assignment statement in an action.

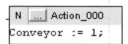

Figure 3-20 A retentive action.

Non-Boolean actions can also be used to call other subroutines. Figure 3-21 shows an example of a transition being used to execute a JSR. This subroutine could be another language or another SFC. JSRs are also commonly used in actions to call other subroutines.

Figure 3-21 ST to call another routine.

Boolean Actions

Boolean actions can also be used. Figure 3-22 shows how a Boolean action is configured. Note the checkmark in the Action Properties screen. The Q bit for this action is used in the logic in Figure 3-23. The Q bit for this action will be true when the action is active.

The use of the Q bit is explained in the table in Figure 3-19. Study the Q bit. If the action is set up as Boolean, the Q bit will be true the entire time the action is active, including the last scan. If the action is non-Boolean, the Q bit will be true while the scan is active until the last scan. The Q bit will be set to false in the last scan. This can be very helpful.

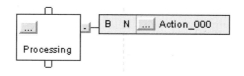

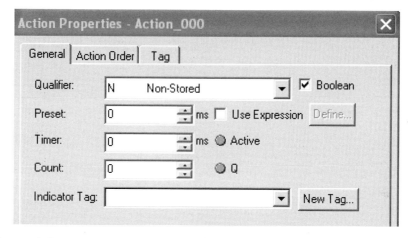

Figure 3-22   A Boolean action. Note that Boolean was checked in the Action Properties screen.

Figure 3-23   A Boolean action being used to call a subroutine named Processing_Cycle.

A Boolean action contains no logic for the action. It is used to set a bit in an action's *member*. To use the actual action, other logic must monitor the bit in the action tag and execute when the bit is set. If you use Boolean actions, you will normally have to manually reset the assignments and instructions that are associated with the action. There is no link between the Boolean action and the logic to perform the action, so the Automatic reset option does not affect Boolean actions.

Figure 3-24 shows an example of a Boolean action. When Step_002 is active, the Boolean Action_007 executes. When the action is active, the Q bit is true for the action. The Q bit is true while the step is active until the last scan when it turns false. In this example, when the step becomes active, the Q bit will be true and Heater_Output will be set to 1. During the last scan of the actions, the Q bit becomes false and Heater_Output is set to 0.

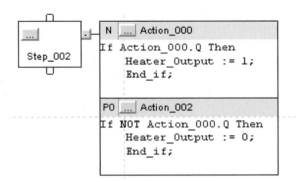

Figure 3-24 Use of the action's Q bit.

A step's Q bit is used in the next example (Figure 3-25). The Q bit is always true if the step is not in its last scan. In the last scan of a step, the Q bit is false (0).

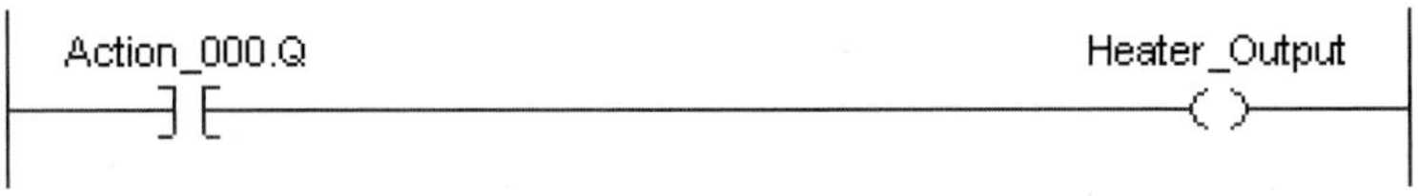

Action_000.Q   Heater_Output

Figure 3-25 Use of an action's Q bit in a ladder diagram.

The Order of Execution for Actions

Actions are executed for a step from top to bottom. The order of the actions for a step can be changed in the step's properties. Figure 3-26 shows how the order of actions can be changed. The programmer simply selects an action and moves it up or down into the desired execution order. Figure 3-27 shows a step and its actions before and after the order of execution was changed.

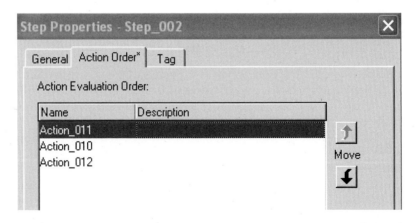

Figure 3-26 The Step Properties setup screen with the Action Evaluation Order chosen.

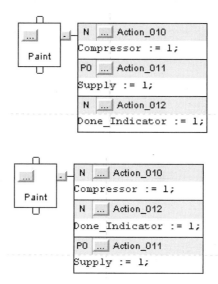

Figure 3-27   Order of execution for actions before and after the order was changed.

Using an Action to Call a Subroutine

Figure 3-28 shows the use of an action to call a subroutine named Temp. Note the tag names between the parentheses. The name of the subroutine is Temp. We will be sending one parameter to the subroutine (Setpoint), and one will be returned from the subroutine (Current_Temp). Figure 3-29 shows the setup screen for the JSR call. The programmer entered Setpoint for the input parameter and Current_Temp for the return parameter. Note that you do not have to send or return values. In this example, the JSR instruction (action) sends the value of the tag Setpoint to the subroutine and the subroutine returns the value of the tag Current_Temp.

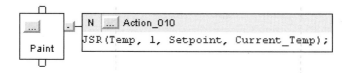

Figure 3-28   An action used to call a subroutine.

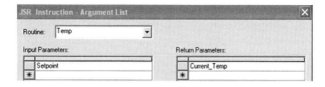

Figure 3-29   Configuration screen for the JSR action in Figure 3-25.

Qualifiers for Actions

Study the table shown in Figure 3-30. There are several action qualifiers that can be used to control how an action executes.

- N is a nonstored action. The N action will stop when the step is deactivated.
- The P1 qualifier executes once when the step is activated.
- S is the stored qualifier. An S action will remain active until a reset action turns off this action.
- A D qualifier causes the action to activate a specific time after a step has been activated and deactivate when a step is deactivated.
- A P qualifier will execute once when the step is deactivated.
- An R qualifier is used to reset (deactivate) a stored step.

If the Action Is to	And	Use	Type
Start when a step is activated.	Stop when the step is deactivated.	N	Nonstored
	Execute only once.	P1	Pulse (Rising edge)
	Stop before the step is deactivated or when the step is deactivated.	L	Time limited
	Stay active until a reset action turns off this action.	S	Stored
	Stay active until a reset action turns off this action or a specific time expires, even if the step is deactivated.	SL	Stored and time limited
Start a specific time after the step is activated and the step is still active.	Stop when the step is deactivated.	D	Time delayed
	Stay active until a Reset action turns off this action.	DS	Delayed and stored
Start a specified time after the step is activated, even if the step is deactivated before this time.	Stay active until a reset action turns off this action.	SD	Stored and time delayed
Execute once when the step is activated.	Execute once when the step is deactivated.	P	Pulse
Start when the step is deactivated.	Execute only once.	PO	Pulse (Falling edge)
Turn off (reset) a stored action. S Stored. SL Stored and time limited. DS Delayed and stored. SD Stored and time delayed.		R	Reset

Figure 3-30   Qualifiers for actions.

TRANSITIONS

Transitions are physical conditions that must happen or change before going to the next step (see Figure 3-31). A transition uses the state of Boolean logic (true or false) to determine whether processing should move to the next step. In Figure 3-32, a tag named Start is used for the Boolean state of the transition.

Transition State	Value	Result
True	1	Go to the next step.
False	0	Continue to execute the current step.

Figure 3-31   A transition table.

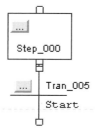

Figure 3-32   A transition. Note that the step will continue to execute until the transition is true.

A transition can be a Boolean or be a JSR instruction to call another routine. Figure 3-33 shows some examples of Boolean transitions. The first example is just a Boolean tag. It could be a sensor's state, for example. If it is true (1), the transition will be true. The second example uses an AND to see if both Boolean tags are true. Both must be true for the transition to be true. The third example uses a Boolean operator (.) and a Boolean tag in an expression. If the expression is evaluated as true, the transition will be true.

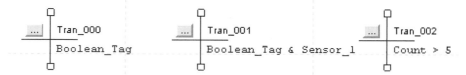

Figure 3-33   Use of Boolean expressions as transitions.

Figure 3-34 shows the use of a JSR instruction in an action to call a subroutine. In this example, the action in Figure 3-34 calls subroutine Heater_On. The Heater_On subroutine is ladder logic that simply turns on an output named Heater_Output (see Figure 3-35).

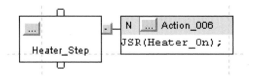

Figure 3-34   Step whose action calls a subroutine named Heater_On.

Figure 3-35   Heater_On subroutine. When this subroutine is called, it turns on Heater_Output.

Figures 3-36 and 3-37 show examples of the use of a JSR instruction in a transition. The JSR instruction in Figure 3-37 is used to call the routine named ST_Decision. The subroutine logic in the ST_Decision routine is used to control the transition state.

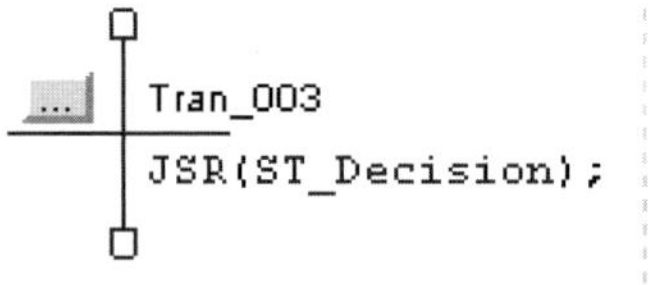

```
    Tran_003
    JSR(ST_Decision);
```

Figure 3-36   This transition calls an ST routine named ST_Decision.

The ST_Decision routine uses ST to decide the state of the transition. Note the use of the End of Transition (EOT) in the last line of the routine to return the state of BOOL-type tag to the transition. An EOT must be used at the end of a subroutine to return a 1 (true) or 0 (false) for the transition.

```
If Sensor_1 & Sensor_2 Then
    Trans_Tag := 1;
Else
    Trans_Tag := 0;
End_if;
EOT(Trans_Tag);
```

Figure 3-37   The ST_Decision routine. An EOT is used to return a value of 0 or 1 to the transition shown in Figure 3-36.

In the example shown in Figure 3-38, a JSR instruction is used to call a subroutine named Part_Done. Part_Done is a ladder logic subroutine. The Part_Done program is shown in Figure 3-39. Sensor_1, State_2, and State_3 must be true to make Boolean_Tag true. If Boolean_Tag is true, the transition will be true. Note the use of the EOT to return the state of Boolean_Tag to the transition.

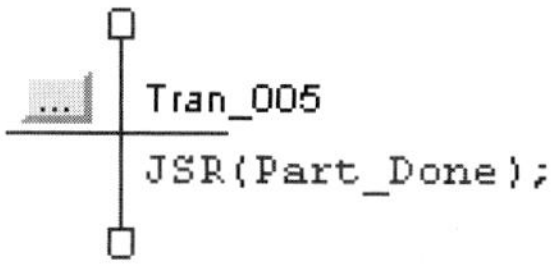

Tran_005

JSR(Part_Done);

Figure 3-38   A transition that calls subroutine Part_Done.

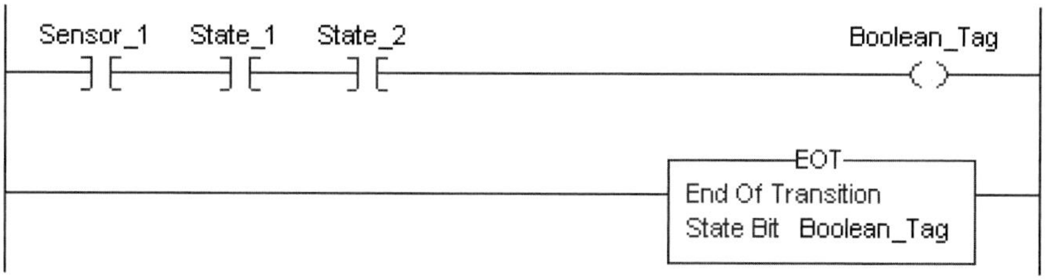

Sensor_1   State_1   State_2   Boolean_Tag

```
—] [———] [———] [————————————————————————————————————————————————————( )—
```

```
                                            ————EOT————
                                            End Of Transition
                                            State Bit  Boolean_Tag
```

Figure 3-39   Subroutine Part_Done. If Sensor_1 AND State_1 AND State_2 are true, the EOT will return a true to the transition and the next step will be executed.

Keeping Outputs on During Multiple Steps

Figure 3-40 shows one method to turn a device on and keep it on for multiple steps. A regular assignment statement could be used to turn a device on as shown in Step_1's action. It will remain on until an assignment statement in an action in a different step turns it off. In this example, it will be turned off by the action in Step_2.

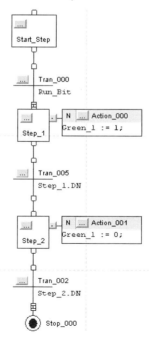

Figure 3-40   Keeping an output on for more than one step.

The example in Figure 3-41 shows the use of a simultaneous branch to keep an output on during multiple steps. This method may make the logic easier to understand.

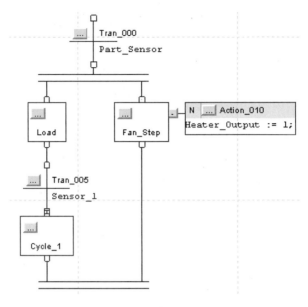

Figure 3-41 A simultaneous branch. Note the branch on the right keeps the Heater_Output on for the whole length of time that the simultaneous branch is active; in fact, it will not turn off at the end of the branch either.

Another method of keeping actions active for multiple steps is by using Stored(S)-type actions. Figure 3-42 shows an example of an S-type action. An S-type action is used to keep the action active. Note that a reset action turns off only the desired action; it does not automatically turn off the devices in that action. You must use another action after the reset action to turn off the device.

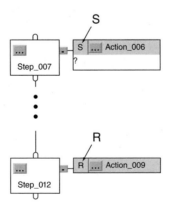

Figure 3-42 Set and reset actions.

Ending an SFC Program

Once an SFC program ends the last step, it does not automatically restart at the first step. If you would like to automatically go back to an earlier step in an SFC program, you would wire the last transition to the top of the step you want to execute next. If you would like to stop after the last scan and wait for a command to restart the SFC program, you would use a stop element. Figure 3-43 shows an SFC program with a Stop element at the end. When the SFC program reaches a stop element, the X bit of the stop element will be set to 1. Stored actions remain active. Execution of the SFC program stops. If a stop element is used in one path of a simultaneous branch, only that path's SFC program will stop scanning; the rest of the SFC program will execute.

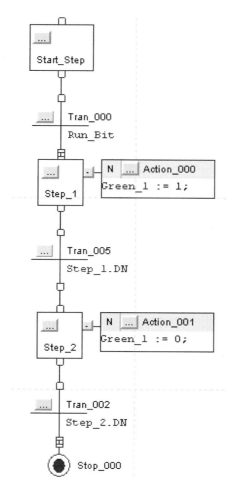

Figure 3-43   Use of a stop element.

Restarting an SFC Program after a Stop

If another SFC program calls the subroutine, it will be reset to the initial step and execute if Automatic reset was chosen for the scan option. Figure 3-44 shows an example of calling another SFC program as a subroutine. If the Programmatic reset option or the Don't scan option was chosen, to restart an SFC program after a stop, you must use an SFC reset (SFR) or logic to clear the X bit of the stop element.

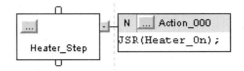

Figure 3-44 Use of a JSR instruction to call and execute another SFC program.

If no other SFC program calls the routine as a subroutine, use an SFR instruction to restart the SFC program at the required step or use logic to clear the X bit of the stop element.

PROGRAMMING A SIMPLE SFC

To begin programming, right click on **MainProgram** and add a new routine (see Figure 3-45). In this example, the routine was given the name SFC_Routine.

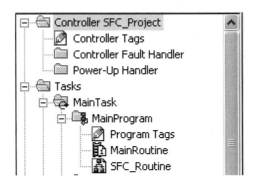

Figure 3-45 SFC Routine added.

Study Figure 3-46. This program has three steps. The program will stay in the Start_step until transition Tran_000 is true. The transition in this case is a Boolean tag named Run_bit. When you try your program, you can force this bit true to move to the second step.

Step_1 is used to turn the green light on for 30 s. The Preset in the step properties must be set to 30,000 ms (30 s). When the step time has reached 30,000, the step's DN bit (Step_1.DN) will be set to 1. This will make the next transition true, and the next step will be executed. Step_2 turns the green light off. Then the program ends at the stop.

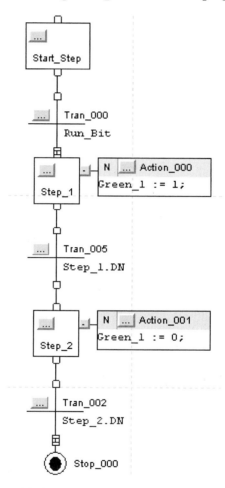

Figure 3-46   SFC program with three steps.

Adding Program Elements

Programming an SFC essentially consists of dragging and dropping program elements (steps and transitions) and then configuring the program elements and adding actions.

To begin the program, select the step and transition icon from the toolbar (see Figure 3-47). Drag it to the programming screen. Next, select another step and transition icon from the toolbar and drag to the programming screen just under the first transition until you see a green dot. When you see the green dot, you can release the mouse button and the second step will be connected to the first transition. Add the third step and transition. Then select the stop icon from the toolbar and drag and drop it below the last transition.

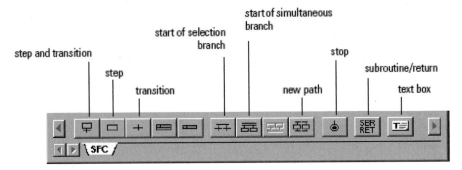

Figure 3-47 The SFC toolbar.

Right click on the first step and select properties or click on the ellipses button on the step to get the properties screen. This is the first step (initial step), so you must choose the Initial checkbox (see Figure 3-48).

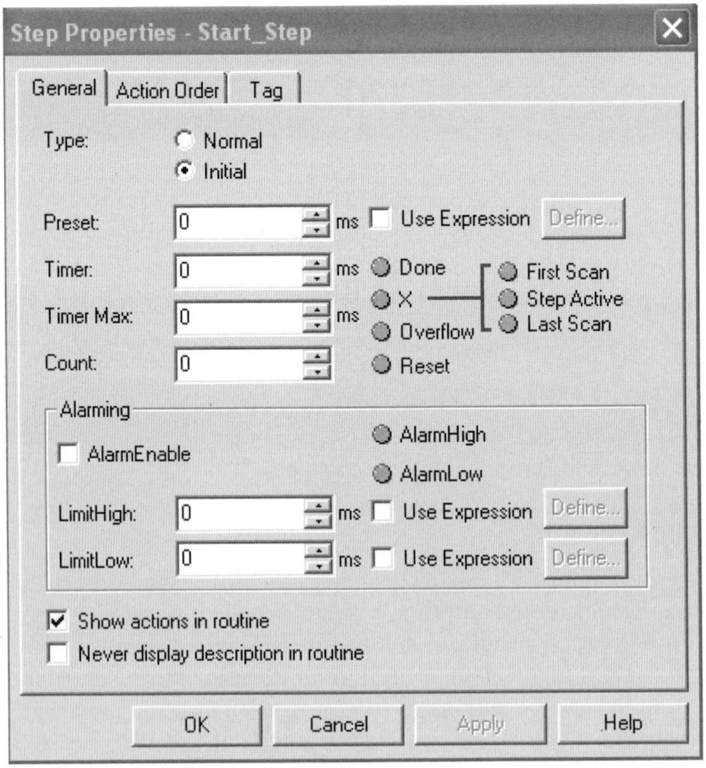

Figure 3-48 Step properties for the initial step.

As shown in Figure 3-49, right click on Step_1, select properties, and enter a Preset of 30,000 ms (30 s). Next you can right click on Step_001 and choose add action. Double click in the bottom of the action, and you can enter your ST for the action as shown in Figure 3-50. Step_2 has one action to turn Green_1 off. This step will not need a preset time. You can add an action for Step_2 and add the action.

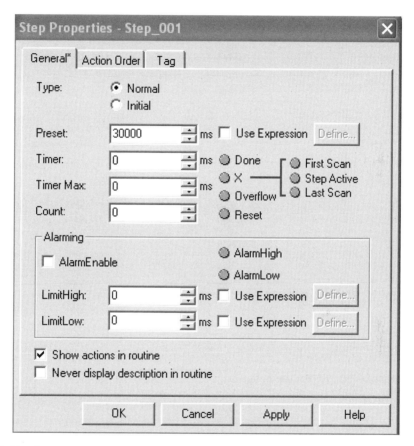

Figure 3-49 Step properties. Note that a Preset of 30,000 ms was used to create a 30-s preset.

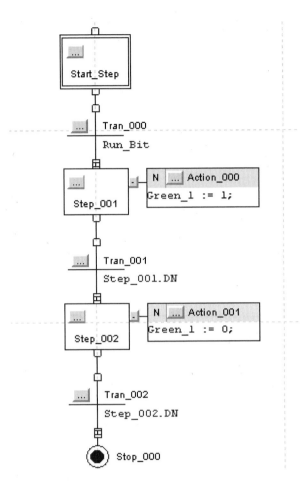

Figure 3-50 ST for the action.

At this point, the program would still have errors because the Green_1 tag has not been created. This program has three step tags, two action tags, one tag in the actions, and three tags in transitions that were created automatically when the elements were created. Right click on the step tag name for the first step. In this example, the tag is named Start_step. Right click on the tag and choose Edit tag to rename the tag. This tag type should be SFC step. Do the same for the other step tags. Next, rename the tags for the transitions. These should be BOOL type. Lastly, create the Green_1 tag. This could be an alias for a real-world output or a bit.

Programming a Simultaneous Branch

Simultaneous branches are programmed, as shown in Figure 3-51. Click on the start of simultaneous branch button on the toolbar (see Figure 3-47), and then drag it to where you want it. Next, add paths to the branches. Click the first step of a path that is to the left of where you want to add the path, and then click on the horizontal line of the simultaneous branch. After you have added the steps, the simultaneous branch can be wired to the preceding transition by clicking the bottom pin of the transition and then the horizontal line of the branch. A green dot will show the valid connection point.

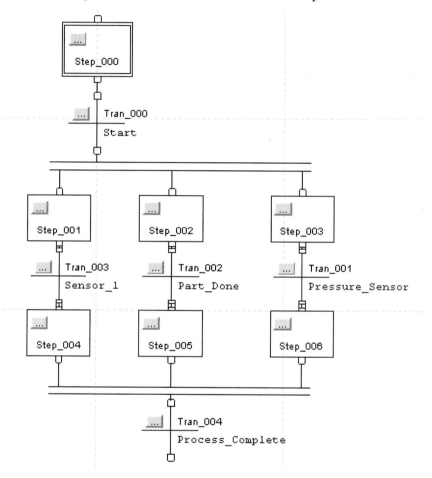

Figure 3-51   A simultaneous branch SFC.

Ending a Simultaneous Branch
To end a simultaneous branch, the last step of each path in the simultaneous branch is selected. Each branch must end with a step, not a transition (you will be connecting the last *step* of each branch to the simultaneous branch end, not transitions). This can be done by clicking and dragging the pointer around all of the desired steps. Or you may click on the first step and then press and hold the shift key down while clicking on the rest of the desired steps. Then you will click on the simultaneous branch end button in the SFC

toolbar. It is located just to the right of the simultaneous branch icon in Figure 3-47. A transition can be added to the branch end.

There is another way to program the end to simultaneous branches. Wire from the connection point on the end of each step to a connection point on the end of the next step, and they will be joined with a simultaneous branch.

Programming a Selection Branch

Figure 3-52 shows a selection branch program. To add a branch, click on the start of selection branch button on the SFC toolbar (see Figure 3-47) and drag the branch to the desired location. To add a path, select (click) the first transition of the path that is to the left of where you want to add the new path. Then click the start of selection branch button. Add the rest of the paths. To wire the selection branch to the preceding step, click the bottom of the step and then click on the horizontal line of the branch. A green dot shows where the connection point is.

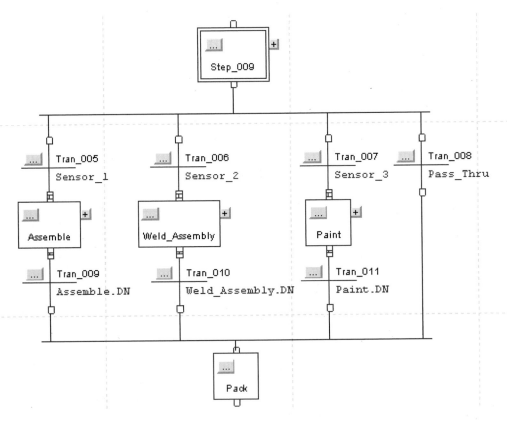

Figure 3-52 Programming a selection branch.

Ending a Selection Branch

Figure 3-51 also shows how to end a selection branch. First you must select all of the last transitions for each path (select the first transition and then hold the shift key down as you select the rest of the transitions). Then click on the end selection branch button. The other way to end a selection branch is to wire the connectors of each of the transitions together, and the end selection branch will be created automatically.

Note that a step follows a selection branch end.

Setting the Priorities for a Selection Branch

A selection branch evaluates the transitions in a selection branch from left to right. The first branch transition that is true will be executed. You may also change the execution priorities for each branch. To change priorities, right click on the horizontal line that starts the selection branch and then choose Set Sequence Priorities. Figure 3-53 shows the priorities screen. Uncheck the Use default priorities check box. Select one sequence at a time and you can move them into the order you would like. Click OK when complete.

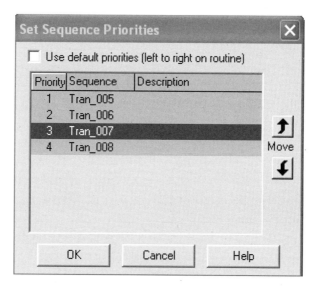

Figure 3-53   Sequence priority screen for a selection branch.

DOCUMENTATION OF SFC PROGRAMS

There are several methods to document an SFC program. Descriptions can be added for tags just like in ladder logic or any of the languages. Comments can be added in ST. Figure 3-54 shows an example of a comment in an action using ST.

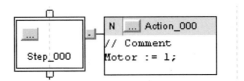

Figure 3-54 Comment added to an action in ST.

Text boxes can be added to program elements to document the program. In Figure 3-55, a text box is added to a step to help explain the purpose of the step. To add a text box, select the text box tool from the toolbar. A text box will appear that you can move around. Select the pushpin on the text box and you can attach it to the desired element by dragging the wire to the element.

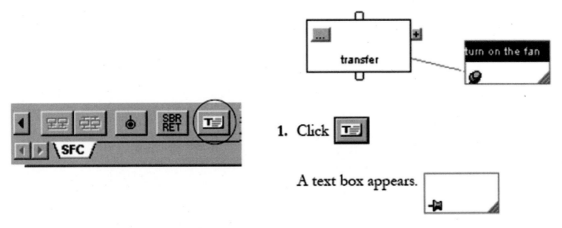

Figure 3-55 Adding text boxes.

QUESTIONS

1. What does SFC stand for?
2. What types of applications is SFC programming best suited for?
3. What is a step?
4. What is an action?
5. How can the order of action execution be changed?
6. What is a transition?
7. On paper, write a start step and a transition. Use a bit named Start for the transition.
8. Write a three-step program on paper. The first step is a start step that will wait until the run bit is true to move to the second step. Step 2 should turn on an output that turns a motor on. The step should execute for 30 s. The last step should turn off the motor output. Explain how the step will be set up to run the motor for 30 s and what will be used for the transition.
9. Write an SFC routine in RSLogix 5000 from question 7 and make it continuously repeat after the run bit is turned on. The output should be on for 30 s and off for 5 s.
10. Explain how the order of execution is determined for selection branches.
11. Explain the following logic:

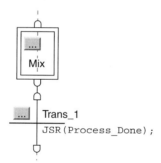

12. Explain the following logic:

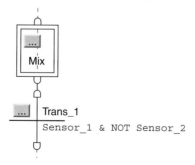

13. Explain the following logic:

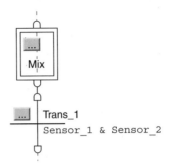

14. Explain the following logic:

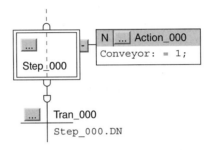

15. Explain the following logic:

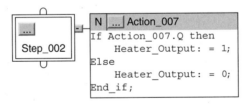

16. Explain the following logic:

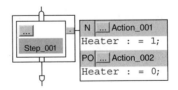

17. Explain the following logic:

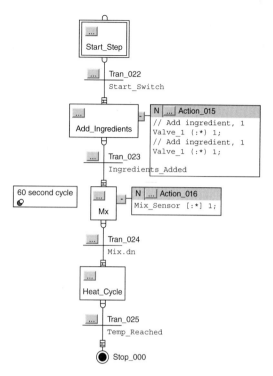

18. Explain the following logic:

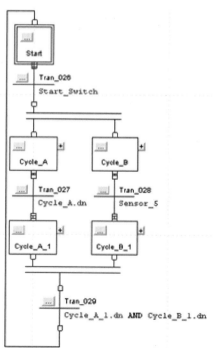

19. How can comments and text boxes be added to an SFC routine?

20. Write logic that uses ladder logic to determine the state of a transition.

21. Write logic in an action to call a subroutine.

22. Develop an SFC routine on paper or in a CLX to accomplish the following:

Step 1	Wait for Start_Switch.
Step 2	Turn a discrete valve named Product_A_Valve on. Turn on the output named HEAT. This step should run until a tag named Level_Fill_Sensor becomes true. When the step is done, the valve should be turned off but the heater output should remain on.
Step 3	Turn on Valve_2, and turn on the output named Mix_Motor. This step should continue until Level_2_Sensor becomes true. When the step ends, you must turn Valve_2 off.
Step 4	This step should continue until the Temp_tag reaches 150°F.
Step 5	Then turn Mix_motor and the heater off. Turn on the drain valve until Tank_Empty_Sensor is true. Run a subroutine named Bottle_Routine.
	Return to Step 1.

23. Write an SFC routine for the following application:

This is a simple heat treat machine application. The operator places a part in a fixture and then pushes the start switch. An inductive heating coil heats the part rapidly to 1500°F. When the temperature reaches 1500, the heating coil turns off and a valve is opened for 10 s to spray water on the part to complete the heat treatment (quench). The operator then removes the part and the sequence can begin again. Note there must be a part present or the sequence should not start. For simplicity, assume that the analog temperature sensor outputs a value that is equal to the actual temperature. The I/O devices for the application are shown below.

I/O	Type	Description
Part_Present_Sensor	Discrete	Sensor used to sense a part in the fixture
Temp_Sensor	Analog	Assume this sensor outputs a value that exactly corresponds to 0–2000°F
Start_Switch	Discrete	Momentary normally open switch
Heating_Coil	Discrete	Discrete output that turns coil on
Quench_Valve	Discrete	Discrete output that turns quench valve on

Function Block Diagram Programming

OBJECTIVES

Upon completion of this chapter, the reader will be able to:

- Explain what is function block programming.
- Explain the types of applications that are appropriate for function block programming.
- Develop function block routines.

INTRODUCTION

Function block diagram (FBD) programming is one of the IEC 61131-3 languages. FBD is a powerful and user-friendly language once you have learned the basics. A function block can take one or more inputs, make decisions or calculations, and then generate one or more outputs. A function block can output information to other function blocks. There are many types of function blocks available to perform various tasks. Function block programming can simplify programming and make a program more understandable. In ControlLogix (CLX), the user develops a function block routine and uses a jump-to-subroutine (JSR) instruction to run the routine from the main routine or another routine.

FBD programming is very useful for applications where there is extensive information/data flow. Process control typically involves more data flow and calculations than discrete manufacturing applications.

ADD Function Block

When you use a function block instruction in a routine, a tag is created for the function block that has several tag members. Figure 4-1 shows an Add instruction in ladder logic and an ADD function block. Note that they are very similar.

The name of the function block (ADD_01) is the default name that is automatically created when you put the ADD function block into the routine. You can use the default tag name or you can change the tag name. The function block tag members are used to store configuration and status information about the instruction. Each function block tag has several tag members that can be used in logic. Figure 4-2 shows the tag and tag members that are created when you use a function block. In this example, 5 tag members were created and all use the name ADD_01 followed by a period (.) and a member name. These tag members can be used in logic.

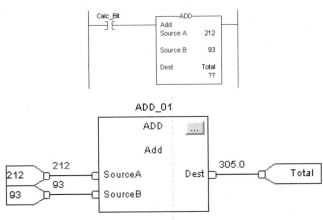

Figure 4-1 Add instruction in ladder logic (top) and an ADD function block (bottom).

The EnableIn tag member (ADD_01.EnableIn) could be used in logic to enable this function block. The ADD_01.SourceA member would be the first of two values to be added. The tag member ADD_01.SourceB would be the second value to be added. Tag member ADD_01.EnableOut can be used to determine whether the result of the instruction is actually output. ADD_01.Dest is the output of the instruction. It is the result of the addition and would be put into the tag member Total. You do not need to use all of these tags. You could simply input a value to the SourceA input and a value to the SourceB input, and the instruction would output the sum of the two values to the Dest output line of the instruction.

Scope: MainProgram	Show: Show All	Sort:
Tag Name	△	**Value**
▶ ⊟ ADD_01		{...}
ADD_01.EnableIn		1
ADD_01.SourceA		0.0
ADD_01.SourceB		0.0
ADD_01.EnableOut		0
ADD_01.Dest		0.0

Figure 4-2 The tag members for an ADD function.

By right clicking on the function block, its parameters can be set (see Figure 4-3). Note that in this example, SourceA, SourceB, and the Dest were enabled. Function block instructions allow the programmer to determine which inputs and outputs will be used. Note that in this example the EnableIn was not checked to be used, but its value is 1, so the instruction is enabled.

	Vis	Name	Value	Type	Description
I	☐	EnableIn	1	BOOL	Enable Input. If False, the...
I	☑	SourceA	0.0	REAL	Source A value
I	☑	SourceB	0.0	REAL	Source B value
O	☐	EnableOut	0	BOOL	Enable Output.
O	☑	Dest	0.0	REAL	Dest value

ADD Properties - ADD_01

Parameters | Tag

Figure 4-3   Properties of an ADD function block.

Function Block Elements

Function blocks are used to take inputs, do some processing, and then provide one or more outputs. There are several methods to get information into and out of a function block. If you want to use tag-type information for an input to a function block, you would use an input reference (IREF). If you want to put the output from a function block into a tag, you would use an output reference (OREF) (see Figure 4-4).

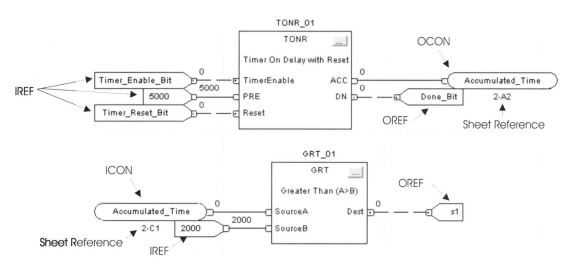

Figure 4-4   IREFs, OREFs, an OCON, and an ICON.

Function blocks can be connected to other function blocks by wiring their outputs to the input of another function block. If there are many function blocks on a sheet, the wiring can make the logic confusing. The other way to connect the output from a function block to the input of a function block is to use output connectors (OCONs) and input connectors (ICONs) (see Figure 4-4). ICONs and OCONs are used as connectors between function blocks. An OCON is an output connector, and an ICON is an input connector. You cannot have an OCON without using an ICON of the same name. The table in Figure 4-5 shows what OCONs, ICONs, IREFs, and OREFs are used for.

Need	Element to Use
To send a value to an output device or a tag.	Output reference (OREF)
To receive a value from an input device or a tag.	Input reference (IREF)
To transfer data between function blocks. Note they can be on the same sheets or on different sheets.	Output connector (OCON) and input connector (ICON)
To send data to several places in a routine.	Single OCON and multiple ICONs

Figure 4-5 Purpose of references and connectors.

When you use an IREF or an OREF, you must create a tag or assign an existing tag to the element. You may use any of the tag data types for an IREF or an OREF.

Order of Execution

The order of execution is controlled by the way elements are wired together and by indicating feedback wires if they are required. The location of a block does not affect the order in which blocks are executed. Figure 4-6 shows an example of a simple FBD and the symbols. Note that the wire type indicates which type of data is being shared. A BOOL value would be a dashed line, and a solid line would indicate a SINT, an INT, a DNT, or a REAL value. If function blocks are not wired together, it does not matter which block executes first as there is no data flow between the blocks. If blocks are wired sequentially, the execution order moves from input to output. The data must be available before a controller can execute a block. In Figure 4-6, the second function block (GRT_02) must execute before the third function block (BAND_02) because the output of the second function block is an input to the third function block.

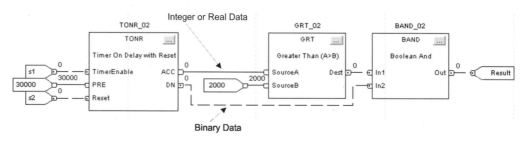

Figure 4-6 Simple FBD.

Figure 4-7 shows two groups of blocks on one sheet. Execution order is important only for the blocks that are wired together.

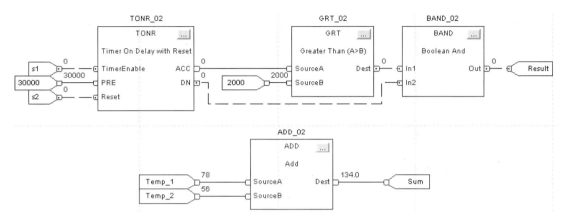

Figure 4-7 Execution of blocks that are not connected.

Feedback

Feedback to a block is done by wiring an output pin from a block to an input pin on the same function block. The input pin would receive the value of the output that was produced on the last scan of the function block. Study Figure 4-8. The figure shows the DN bit (TONR_02.DN) used to reset the timer.

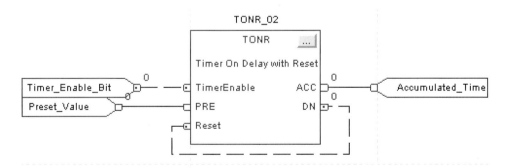

Figure 4-8 Feedback loop.

A controller cannot determine which function block to execute first for function blocks that are in a loop. The programmer must identify which block should be executed first by marking an input wire with the Assume Data Available Marker (see Figure 4-9). The arrow indicates that this data serves as the input to the first function block in the loop. Only one input to a function block in a loop should be marked. To add the Assume Data Available Marker, select the wire, right click the mouse, and select the Assume Data Available choice.

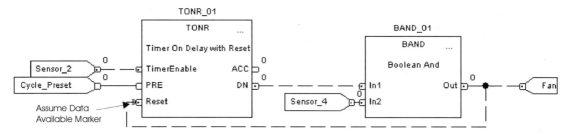

Figure 4-9 Assume Data Available Marker.

Figure 4-10 shows that if there is more than one connection between function blocks, either all must be marked with the Assume Data Available Marker or none must be marked. The top example in Figure 4-10 is incorrect. The bottom example is correct.

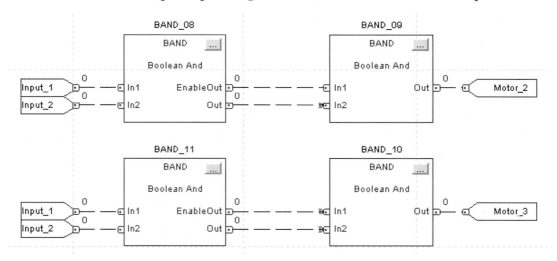

Figure 4-10 Use of the Assume Data Available Marker.

The Assume Data Available Marker can be used to create a one-scan delay between blocks (see Figure 4-11). In this figure, the first block is executed and then the second block uses the data that was generated in the previous scan of the function block routine.

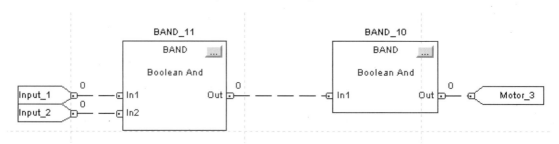

Figure 4-11 Use of Assume Data Available Marker to create a one-scan delay.

IREFs are used to provide input data to a function block instruction. The data in an IREF is latched (won't change) during the function block scan. IREF data is updated at the beginning of a function block scan.

The following is a summary of the execution of a function block scan:

1. The processor latches the data in all IREFs.
2. The processor executes the function blocks in the order determined by their wiring.
3. The processor writes outputs to the OREFs.

Connectors

ICONs and OCONs are used to transfer data between output and input pins (see Figure 4-12). They can be used to pass information between function blocks instead of wires when the elements you want to connect are on different sheets, when a wire might be hard to route on a sheet, when you want to provide the data to several points in a routine, or when you wish to pass data to another sheet of FBD. Note from Figure 4-12 that the output Accumulated_Time from the TONR_01 function block is used as the input Accumulated_Time to the GRT_01 function block. Note also that if SourceA (Accumulated_Time) were greater than SourceB (2000), the output (Dest) would be true. Accumulated_Time is not greater than 200 ms in this example, so the output is false. Note that the input lines to the GRT function block are solid and the output line is dashed. The dashed line means that the output is a discrete value (1 or 0).

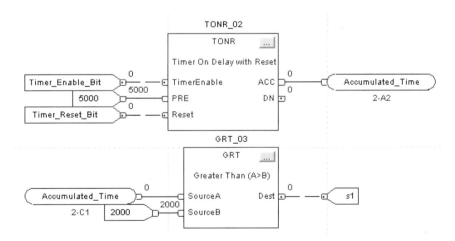

Figure 4-12 Use of connectors instead of wires.

Using Connectors
Each OCON must have a unique name. Connector names follow tag name rules, although they are not tags. Each OCON must also have at least one corresponding ICON. In other words, there must be at least one ICON with the same name as the OCON.

Multiple ICONS can be used for the same OCON. This enables you to use an output value (OCON) in multiple places in your routine as an ICON.

MATHEMATICAL FUNCTION BLOCKS

There are many types of function blocks available. Mathematical function blocks are one type. Let's examine a few function blocks to see how a typical function block works.

ADD Function Block

An ADD function block can be used to add two numbers or the values of tags or numbers. Figure 4-13 shows an example of the use of an ADD function block. In this example, two constants (numbers) were used as inputs. Here, 212 was added to 93 and the result was put into the OREF tag named Total. You can see that the result (305.0) is also shown above the output pin.

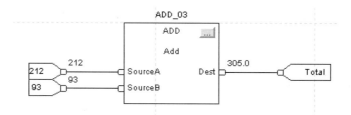

Figure 4-13   An ADD function block.

SUB Function Block

A SUB function block can be used to subtract two numbers or the values of tags or numbers. Figure 4-14 shows an example of the use of a SUB function block. In this example, two constants (numbers) were used as inputs. Here, 93 was subtracted from 212 and the result was put into the OREF tag named Total. You can see that the result (119.0) is also shown above the output pin.

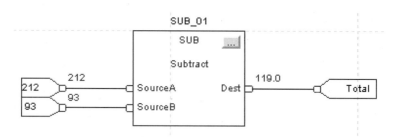

Figure 4-14   A SUB function block.

MUL Function Block

A MUL function block is shown in Figure 4-15. There are two inputs and one output. In this example, 212 was multiplied by 93 and the answer (19,716.0) was output to the OREF tag named Total.

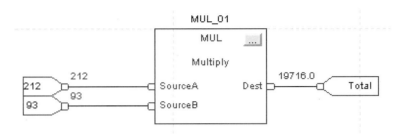

Figure 4-15   A MUL function block.

DIV Function Block

A DIV function block is shown in Figure 4-16. There are two inputs and one output. In this example, 212 was divided by 93 and the answer (2.2795699) was output to the OREF tag named Total.

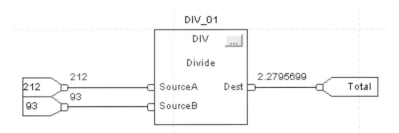

Figure 4-16   A DIV function block.

Boolean AND Function Block

Figure 4-17 shows an example of ladder logic to create Boolean AND logic and also a function block Boolean AND (BAND). A BAND function block can be used to compare two or more discrete inputs (see Figure 4-17). If all are true, the discrete output of the function block will be true. If any or all are false, the output will be false.

This is a very useful instruction. There are many times in an application when we need to do something if exact input conditions are met. For example, if Sensor_1 is true AND Sensor_2 is true AND Sensor_3 is true AND Sensor_4 is true and we want to output a true from the instruction, BAND is the perfect instruction.

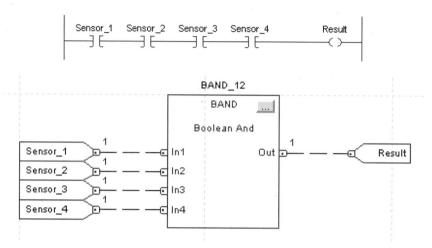

Figure 4-17 Ladder logic to create Boolean AND logic (top) and a BAND function block (bottom).

Boolean OR (BOR) Function Block

A BOR function block can be used to compare two or more discrete inputs. If any are true, the discrete output of the function block will be true. Figure 4-18 shows an example of how a Boolean OR is created in ladder logic and also a BOR function block. Note that there are four discrete inputs in this example. If one, some, or all of the four inputs are true, the output will be set to true. In this example, only input 3 is true and the output is set to true. Note that if you right click on a BOR function block, you can reduce or increase the number of inputs.

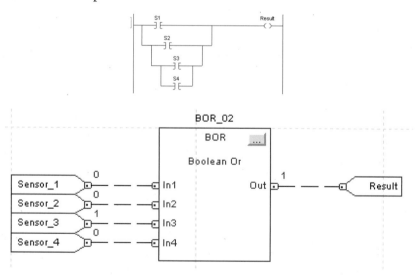

Figure 4-18 A Boolean OR (top) and a BOR function block (bottom).

Figure 4-19 shows additional Compute/Math function blocks that are available. These function blocks include add, subtract, multiple, divide, modulo, square root, negate, and absolute. The MOD instruction is used to find the remainder of a division. The NEG instruction is used to change the sign of the Source and places the result in the Dest. The absolute instruction (ABS) takes the absolute (positive) value of the Source and places the result in the Dest.

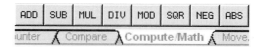

Figure 4-19   Compute/Math function block instructions.

TRIGONOMETRIC FUNCTION BLOCKS

Figure 4-20 shows trigonometric function blocks. Trigonometric function blocks include sine, cosine, tangent, arc sine, arc cosine, and arc tangent.

Figure 4-20   Trigonometric function block instructions.

STATISTICAL FUNCTION BLOCKS

Figure 4-21 shows the statistical function blocks that are available. A MAVE instruction is a moving average instruction. The MSTD instruction can be used to calculate the moving standard deviation for a process. The MIN instruction is actually a MINC instruction. It means minimum capture. The MINC instruction finds the minimum of an input signal to the instruction over time. The MAX instruction is actually the MAXC instruction. The MAXC instruction finds the maximum of an input signal over time.

Figure 4-21   Statistical function block instructions.

Moving Average (MAVE) Instruction

The MAVE instruction calculates an average value, over time, for the In (input) signal. This instruction optionally supports user-specified weights. It is available in function block and structured text (ST) programming. An example of the use of this instruction is to monitor the size of the product that is being made. This instruction could look at a moving average so that the correct adjustment could be made on the basis of the average size of a number of products that are made rather than just the last one made. This can make adjustments more accurate.

Initializing the Averaging Algorithm

Certain conditions, such as instruction first scan and instruction first run, require the instruction to initialize the moving average algorithm. Figure 4-22 shows an example of a MAVE instruction.

Each scan, the instruction places the input value from the In_Value tag in the StorageArray named Values. The most current input is put in the first element (Values[0] in this example) of the array named Values. The instruction calculates the average of the values in the StorageArray, optionally using the weight values in array weight, and places the result in Out. Note that a new value is input every scan that this instruction is true.

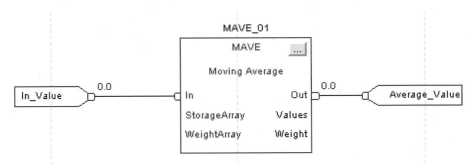

Figure 4-22   Use of a MAVE instruction.

Minimum Capture (MINC) Instruction

The MINC instruction finds the minimum of the input signal over time (see Figure 4-23). A good example of this might be to record the lowest temperature in a process during one day of operation. This instruction is available in function block and ST programming. The parameters for a MINC instruction are shown in Figure 4-24.

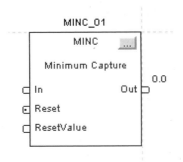

Figure 4-23   A MINC function block.

Inputs/Outputs	Data Type	Description
EnableIn	BOOL	If the Enable Input (EnableIn) is cleared, the instruction does not execute and outputs are not updated.
In	REAL	This is the analog signal input to the instruction. Any float is valid.
Reset	BOOL	This is a request to reset the control algorithm. This instruction sets Out = ResetValue as long as Reset is set. Any float is valid.
ResetValue	REAL	This is the reset value for the instruction. This instruction sets Out = ResetValue as long as Reset is set. Any float is valid.
EnableOut	BOOL	Enable output.
Out	REAL	This is the calculated output of the algorithm.

Figure 4-24   Parameters for a MINC instruction.

There is also a maximum capture (MAXC) instruction available, to capture a maximum value from an input. An example of its use is to record the highest temperature during a day of production.

MATHEMATICAL CONVERSION FUNCTION BLOCK INSTRUCTIONS

Figure 4-25 shows some mathematical conversion function blocks. A DEG instruction can be used to convert radians to degrees. A RAD instruction is used to convert degrees to radians. A TOD instruction can be used to convert an integer to a BCD value. An FRD (convert to integer) instruction can convert a BCD value to an integer. A truncate (TRN) instruction is used to truncate an integer or a real value.

Figure 4-25   Mathematical conversion function block instructions.

Scale (SCL) Instruction

The SCL instruction converts an unscaled input value to a floating-point value in engineering units. These are very useful for converting the counts from an analog value to a number that makes more sense to an operator, for example, scaling the counts from an encoder on a motor to an actual speed in RPMs. Figure 4-26 shows the use of an SCL instruction. In this example, the input raw values will be between 0 and 600. The SCL instruction will scale the input to a value between 0 and 60.

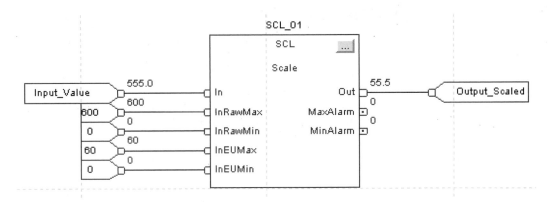

Figure 4-26 An SCL function block instruction.

FUNCTION BLOCK TIMERS

Figure 4-27 shows a timer on delay with reset (TONR) function block. It should look fairly familiar. It has the same basic inputs, parameters, and outputs as a TONR ladder logic timer. This timer was named TONR_01. TONR_01 is the default name and it can be changed. The TimerEnable input is used to enable the timer. In this example, an IREF uses a tag named Start to enable the timer. A constant value (30,000) was used in an IREF for the PRE value. The time base for CLX timers is milliseconds, so this timer would be a 30-s timer. The accumulated time of the timer is being output to an OREF tag named ACC. The timer's DN bit (TONR_01.DN) is being used to reset the timer.

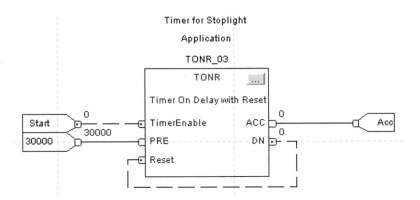

Figure 4-27 A TONR function block. Note that the DN bit was wired to the Reset input of the timer to automatically reset the timer when the DN bit becomes true. Note also that the connection is a dashed line. This means the connection passes a discrete value.

FUNCTION BLOCK COUNTERS

Figure 4-28 shows a function block counter. This counter can be used to count up and to count down.

Note that there is a count-up enable input and a count-down enable input. There is also a preset input (PRE) and a Reset input. The counter outputs include the present count (ACC) and a DN bit. The DN bit is set if the ACC is equal to the PRE.

Imagine a manufacturing cell where a component enters the cell to be worked on. As it enters the cell, we might want to add it to the count of parts ready to be worked on. We could use a sensor (Part_In_Sensor in this example) to sense the part coming in as an input to the count-up input of the counter. As a part is finished and leaves the cell, we could have a sensor (Part_Out_Sensor) sense it leaving and use it as an input to the count-down input of the counter. The counter's accumulated value would always contain the number of parts actually in the cell. If we set the PRE value to 2, we could use the DN bit to warn the operator when there are only two parts left in the cell.

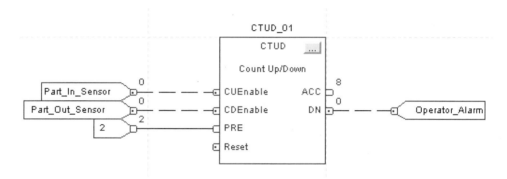

Figure 4-28   A count-up and count-down counter.

Figure 4-29 shows the timer and counter instructions that are available in function block programming. The TONR instruction is a nonretentive timer that accumulates time when TimerEnable is set. The TOFR instruction is a nonretentive timer that accumulates time when TimerEnable is cleared. The RTOR instruction is a retentive timer that accumulates time when TimerEnable is set. The CTUD instruction counts up by 1 when CUEnable transitions from clear to set. The instruction counts down by 1 when CDEnable transitions from clear to set.

Figure 4-29   Function block timers and counters.

PROGRAMMING FUNCTION BLOCK ROUTINES

To start a function block program, right click on the **MainProgram** (see Figure 4-30). Then you will add a new routine. Choose Function Block for the type of routine and give it a name. In this example, the routine was given the name Stop_light_FB. It is shown below the MainProgram in the Controller Organizer in Figure 4-30.

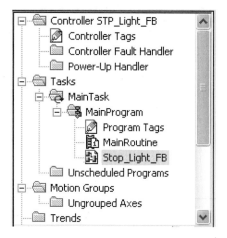

Figure 4-30   Program list in RSLogix 5000. Note the routine named Stop_light_FB. The ICON shows that it is a function block routine.

The next example uses timer (TONR) and limit (LIM) function block instructions. Figure 4-31 shows the use of a TONR timer function block. The TONR has a preset of 30,000 ms (30 s). It has accumulated a count of 28,385 at this point in time (about 28 s).

The LIM function block will be used to check to see if a value is between a low and a high limit. If the input to the LIM function block Test input is between the LowLimit and HighLimit, the output from the LIM function block will be true. In this example, the output will be true only if the input is between 0 and 10,000. In this example, the output from the LIM function block is a real-world output.

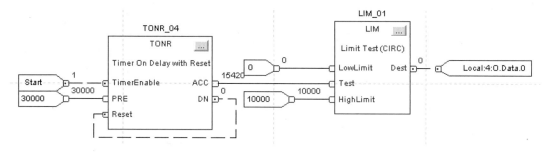

Figure 4-31   Two function blocks wired together.

Study Figure 4-32. This is the same application as shown in Figure 4-31, but programmed slightly differently. The two function blocks in this example are not wired together. An OCON and an ICON are used to pass information between the two function blocks. Note that the input to the LIM instruction is an ICON named ACC in this example. ICON ACC gets its data from the OCON output from the TONR function block. Note that these two function blocks are not connected but work together. The TONR output OCON (ACC) provides input to the TEST input of the LIM through the ICON (ACC). If ACC is between 0 and 10,000, the LIM function block's output will be set to true. The numbers and letters under the ACC output show the page and are of the page where this output is used. Note that the output is an OCON named ACC. This application would perform exactly as the one shown in Figure 4-31. Note also that the use of an OCON and an ICON would enable these two function blocks to be on different sheets of the function block routine.

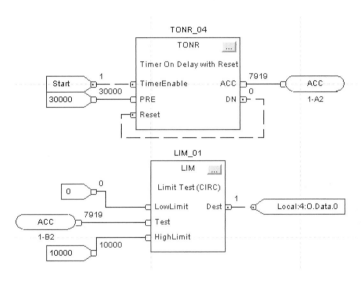

Figure 4-32   TONR and LIM function blocks.

In RSLogix 5000, a function block routine can be broken into multiple sheets. Sheets are like separate pages of a program (see Figure 4-33). This helps you organize your program and make it easier to understand. Sheets do not affect the order in which the function blocks execute. When a function block routine executes, all sheets execute. It is a good idea to use one sheet for each device that is to be programmed. Figure 4-33 shows four sheets (pages) of an FBD program. In this example, each sheet controls one device. Note that this is about as simple as it gets. Normally, there might be multiple function blocks on a page to perform different but related tasks.

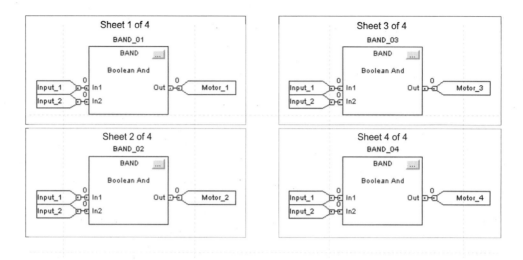

Figure 4-33 Multiple sheets in a function block routine.

ADDITIONAL FUNCTION BLOCKS

Select (SEL) Function Block

The SEL function block uses a digital input to select one of two inputs. This instruction is available only in function block programming. An example is shown in Figure 4-34.

The SEL function block selects In1 or In2 on the basis of SelectorIn. If SelectorIn is set, the instruction sets Out = In2 (see Figure 4-34). If SelectorIn is cleared, the instruction sets Out = In1. In this example, 0 was input to SelectorIn so the value at In1 (150) is output to the OREF named Result.

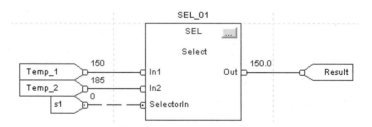

Figure 4-34 A SEL function block.

PROGRAM/OPERATOR CONTROL OF FUNCTION BLOCKS

Several function blocks support program/operator control. Program/operator control enables the programmer to control these instructions alternatively from a program or from an operator interface device. When an instruction is in program control, the instruction is controlled by the program inputs to the instruction. When an instruction is under operator control, the instruction is controlled by the operator inputs to the instruction. Program or operator control is determined by the inputs shown in the table in Figure 4-35.

Input	Description
.ProgProgReq	A program request to go to program control
.ProgOperReq	A program request to go to operator control
.OperProgReq	An operator request to go to program control
.OperOperReq	An operator request to go to operator control

Figure 4-35 Program/operator control inputs and options.

If both ProgProgReq and ProgOperReq are set, the instruction will be under operator control.

You can determine whether the instruction is in program or operator control by looking at the ProgOper output. If ProgOper is set (1), the instruction is in program control. If ProgOper is set (0), the instruction is in operator control.

Program request inputs take precedence over operator request inputs. This enables the user to use the ProgProgReq and ProgOperReq inputs to lock an instruction in the desired mode. For example, assume you always want an instruction to operate in operator mode. You do not want the program to control the running or stopping of the instruction. To do this, you would input 1 into the ProgOperReq. This would prevent the operator from putting the instruction into program control by setting the OperProgReq from an operator input device. Let's examine one instruction that utilizes program/operator control.

Enhanced Select (ESEL) Function Block

The ESEL instruction (see Figure 4-36) lets you select one of as many as 6 inputs or the highest, lowest, median, or average of the inputs and send the result to the output. The SelectorMode input (see Figure 4-37) value determines whether the instruction will select the highest, lowest, median, or average of the inputs. In the example shown in Figure 4-36, 4 is the input to the SelectorMode input, so the average of the input values will be output by the instruction. There are also inputs to determine if the instruction is under program or operator control. This instruction is available in function block and in ST.

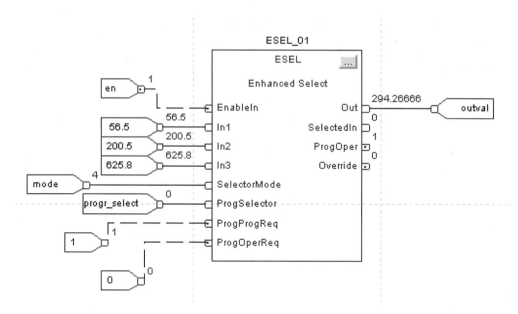

Figure 4-36 An ESEL instruction.

Value	Description
0	Manual select
1	High select
2	Low select
3	Median select
4	Average select

Figure 4-37 Selector modes.

Switching between Program Control and Operator Control

The following list states how the ESEL instruction changes between program control and operator control.

1. You can lock the instruction in operator control mode by leaving ProgOperReq set.
2. You can lock the instruction in program control mode by leaving ProgProgReq set while ProgOperReq is cleared.

Multiplex (MUX) Instruction

The MUX instruction can be used to select 1 of 8 inputs to send to the output on the basis of the selector input. On the basis of the selector value, the MUX instruction sets Out equal to 1 of the 8 inputs. The number of inputs can be reduced. An example of the use of this might be a process where we have different potential temperatures, depending on the product that needs to be produced.

Figure 4-38 shows the use of a MUX instruction to choose which one of 8 input values should be sent to the output. This MUX instruction selects between In1, In2, In3, In4, In5, In6, In7, and In8, on the basis of the selector. The instruction sets Out = In, which becomes an input parameter for MUX_01. For example, if Select_Value = 2 (value into the selector input), the instruction sets Out = Analog_Input2, 6.7 in this example.

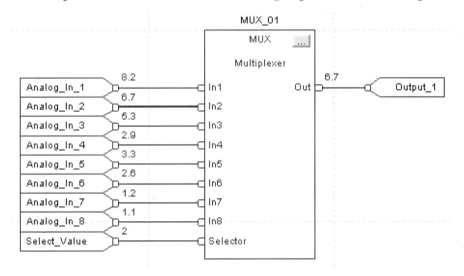

Figure 4-38   A MUX instruction.

This chapter has covered only a small number of the available function block instructions. If you become comfortable with these, you will readily learn others that you need to use. You can bet that there is an instruction or a combination of instructions that will solve any application need you have. If it is a recurring need, you can even create your own Add-On instruction for future use.

QUESTIONS

1. Function block programming is one of the languages that IEC 61131-3 specifies. (True or False)
2. What does the acronym IREF stand for?
3. What does the acronym OREF stand for?
4. Can an IREF be a tag? A number?
5. Can an OREF be a tag?
6. What does the acronym ICON stand for?
7. What does the acronym OCON stand for?
8. What is the difference between an OREF and an OCON?
9. What are ICONs and OCONS used for?

10. What is the Assume Data Available indicator used for?

11. What is a sheet in a function block routine?

12. What are two ways you could get information from an instruction on one sheet to an instruction on another sheet?

13. Thoroughly explain the following logic. Make sure you explain the types of inputs and outputs to each instruction as well as what the logic does.

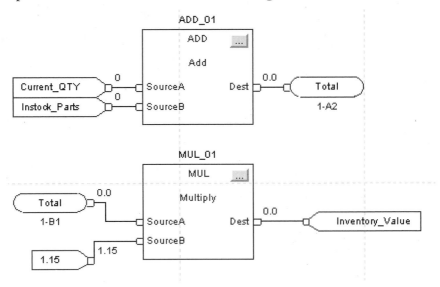

14. Thoroughly explain each instruction and the following logic. Make sure you explain the types of inputs and outputs to each instruction as well as what the logic does.

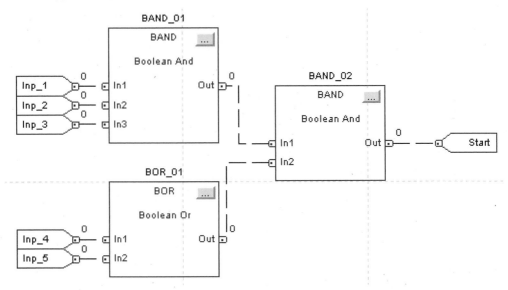

15. Thoroughly explain each instruction and the following logic. Make sure you explain the types of inputs and outputs to each instruction as well as what the logic does.

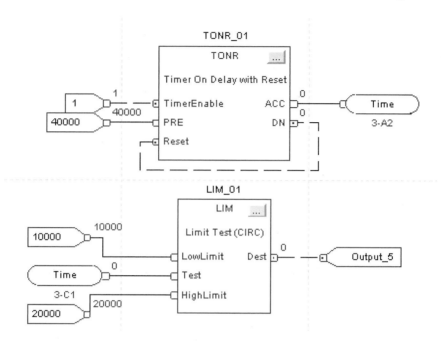

16. Thoroughly explain each instruction and the following logic. Make sure you explain the types of inputs and outputs to each instruction as well as what the logic does.

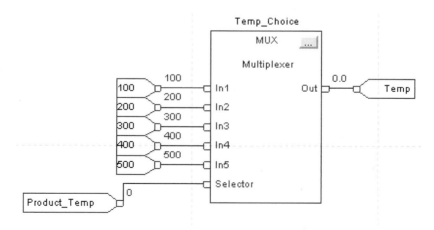

17. Thoroughly explain each instruction and the following logic. Make sure you explain the types of inputs and outputs to each instruction as well as what the logic does.

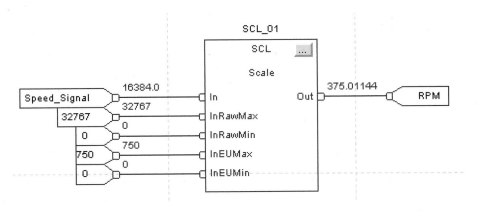

18. Write a function block routine for the following application. You may do it on paper or on a controller.

This is a simple heat treat machine application. The operator places a part in a fixture and then pushes the start switch. An inductive heating coil heats the part rapidly to 1500 degrees Fahrenheit. When the temperature reaches 1500, turn the coil off and open the quench valve, which will spray water on the part for 10 s to complete the heat treatment (quench). The operator then removes the part and the sequence can begin again. Note there must be a part present or the sequence should not start. Note: you may want to use a small ladder diagram program for the start/stop logic. It will simplify the task.

I/O	Type	Description
Part_Present_Sensor	Discrete	Sensor used to sense a part in the fixture
Temp_Sensor	Analog	Assume this sensor outputs 0–2000 degrees Fahrenheit. To keep it easy assume that the sensor is analog and will output the number 1500 when the temperature reaches 1500
Start_Switch	Discrete	Momentary normally open switch
Heating_Coil	Discrete	Discrete output that turns coil on
Quench_Valve	Discrete	Discrete output that turns quench valve on

Add-On Instructions

OBJECTIVES

Upon completion of this chapter, the reader will be able to:

- Explain the benefits and uses of Add-On instructions.
- Utilize various languages to develop Add-On instructions.
- Develop Add-On instructions with complete documentation.

ADD-ON INSTRUCTIONS

Add-On instructions are custom instructions you can create yourself. Add-On instructions can be used to create new instructions for sets of commonly used logic. Add-On instructions can be developed in ladder logic, structured text, or function block. If you develop an Add-On instruction in one language, function block for example, ControlLogix automatically creates the instruction in ladder logic and structured text for you. As you develop the instruction, you are automatically developing the documentation for it.

The following are some of the benefits of Add-On instructions:

- If there is logic that is used multiple times in the same project or in different projects, it may make sense to put the logic inside an Add-On instruction to make the logic modular and easier to reuse. It may also make the logic easier to understand, hiding some of the complexity behind the code.
- Add-On instructions allow a programmer to reuse the work he or she has invested to develop special logic (algorithms).
- Add-On instructions can provide consistency between projects by enabling the reuse of commonly used control algorithms.

- Add-On instructions allow the programmer to put complicated algorithms inside an Add-On instruction and then provide an easier-to-understand interface by making only essential parameters visible.
- The use of Add-On instructions reduces the time required to document the instruction by automatically generating instruction help.
- The proprietary code you develop can be put inside the Add-On instruction, and Source Protection can be used to prevent others from viewing or changing your code. This can help keep your code proprietary. This is very important for companies that manufacture and sell automation equipment.
- Once an Add-On instruction is defined in a project, it behaves like the standard instructions already available in the RSLogix software. Add-On instructions appear on the instruction toolbar.

DEVELOPING AN ADD-ON INSTRUCTION

In this example, we will create an Add-On instruction that will take a bit count that represents the RPM of a motor as an input, convert the bit count to an actual RPM value, and control two output bits. One of the bits will be set to a 1 if the RPM is in the correct speed range (0–900 RPM). The other output bit will be set to a 1 if the speed is too high (900–1200). The logic for the Add-On instruction that will be developed in this example is shown in Figure 5-1.

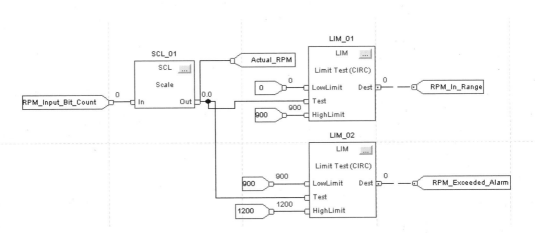

Figure 5-1 Logic for the Add-On instruction example.

To create an Add-On instruction, right click on **Add-On Instructions** in the controller organizer as shown in Figure 5-2.

After you have chosen **New Add-On Instruction**, the General screen, shown in Figure 5-3, will appear. Note the tabs on the top of the screen: General, Parameters,

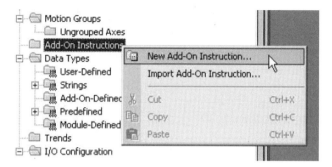

Figure 5-2 Creating a new Add-On instruction.

Local Tags, Scan Modes, Change History, and Help. Also note the **Logic** button on the lower left. The logic can be entered at any time by clicking on the **Logic** button. The logic can be developed in ladder logic, structured text, or function block. There are advantages to creating the logic first. The parameters and local tags are created while you create the logic.

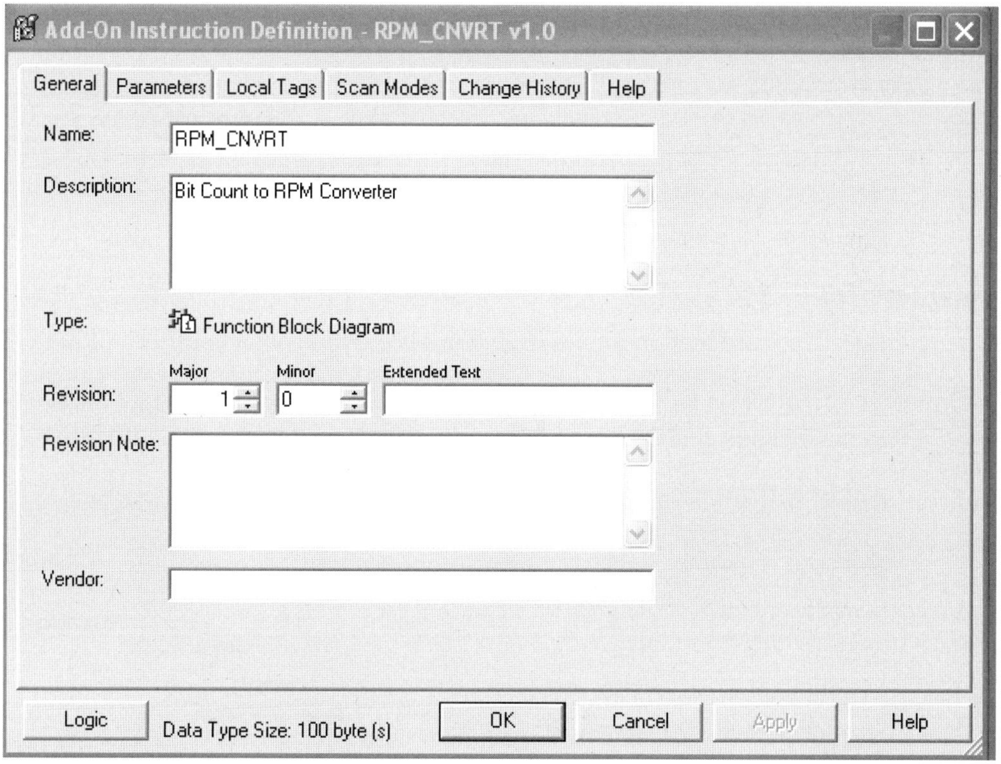

Figure 5-3 The General information screen.

General Tab

Figure 5-3 shows the information that can be entered in the General tab, which can be used to update the information. Also note that the description, revision, revision note, and vendor information are copied into the custom help for the instruction. The programmer is responsible for defining how the revision level is used and when it is updated. Revision levels are not automatically managed by the software. Note that the name of the Add-On instruction for this example is RPM_CNVRT.

Parameters Tab

The Parameters define the instruction interface and also how the instruction appears when used in logic. The Parameter order that you develop defines the order that the Parameters appear on the instruction. Figure 5-4 shows the Parameter input screen. The parameters that are used in the logic for RPM_CNVRT are shown. Note also that in this example the logic was created first. As the input and output references were created, they were automatically added to the parameters list.

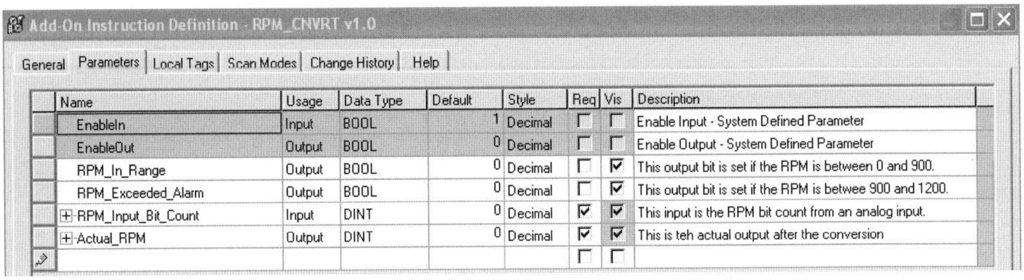

Name	Usage	Data Type	Default	Style	Req	Vis	Description
EnableIn	Input	BOOL	1	Decimal	□	□	Enable Input - System Defined Parameter
EnableOut	Output	BOOL	0	Decimal	□	□	Enable Output - System Defined Parameter
RPM_In_Range	Output	BOOL	0	Decimal	□	☑	This output bit is set if the RPM is between 0 and 900.
RPM_Exceeded_Alarm	Output	BOOL	0	Decimal	□	☑	This output bit is set if the RPM is betwee 900 and 1200.
⊞ RPM_Input_Bit_Count	Input	DINT	0	Decimal	☑	☑	This input is the RPM bit count from an analog input.
⊞ Actual_RPM	Output	DINT	0	Decimal	☑	☑	This is teh actual output after the conversion
					□	□	

Figure 5-4 Parameters screen.

Local Tags Tab

Local Tags are hidden members and are not visible outside of the Add-On instruction. They cannot be referenced by other programs or routines. They are private to the instruction. This can be very important. It can make the instruction simpler and easier to understand. It also may be of benefit to hide proprietary logic. Figure 5-5 shows the

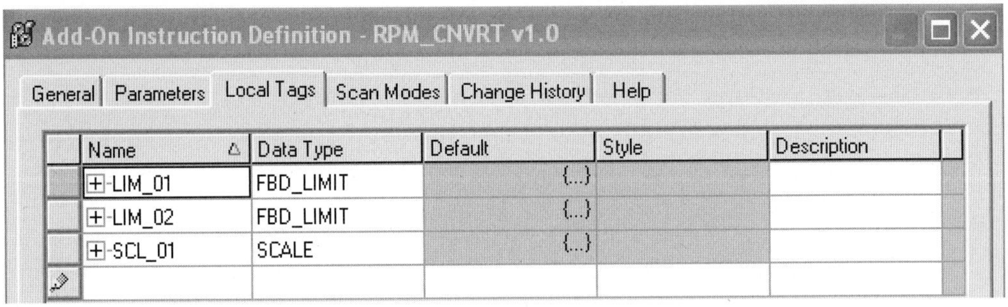

Name	△	Data Type	Default	Style	Description
⊞-LIM_01		FBD_LIMIT	{...}		
⊞-LIM_02		FBD_LIMIT	{...}		
⊞-SCL_01		SCALE	{...}		

Figure 5-5 Local Tag screen.

Local Tag screen. As the logic was created for the instruction, the Local Tags were automatically added to the Local Tag list.

Scan Modes Tab

The Scan Modes tab (see Figure 5-6) enables you to define additional routines for Scan mode behavior.

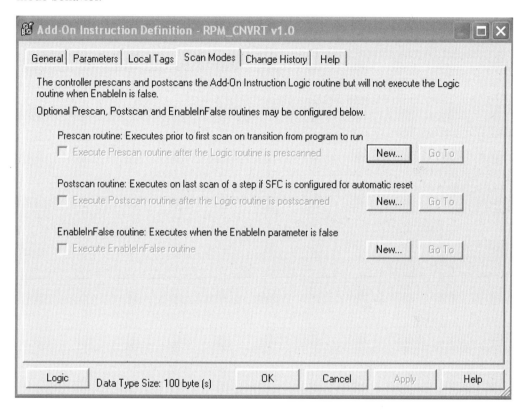

Figure 5-6   Scan Modes screen.

There are three types available: Prescan, Postscan, and EnableInFalse. The Prescan would execute prior to the first scan on a transition from Program to Run. Prescan routines for Add-On instructions are optional. A Prescan routine for an Add-On instruction provides a way for an Add-On instruction to define additional behavior for Prescan mode. Generally, if a Prescan routine is defined and enabled, it executes after the primary Logic routine executes in Prescan mode. Prescan mode could be used to initialize tag values to some known or predefined state prior to execution.

The Postscan would execute on the last scan of a step if a Sequential Function Chart (SFC) has been configured for automatic reset. With the Automatic Reset option set, when an Add-On instruction is called by logic in an SFC action or an Add-On instruction call resides in a routine called by a jump-to-subroutine (JSR) from an SFC action, the Add-On instruction will execute in Postscan mode. The primary Logic routine of the Add-On instruction executes in Postscan mode. Then, if it is defined and enabled, the Postscan routine for the Add-On instruction executes. This is useful to reset internal states, status values, or to de-energize the Add-On instruction outputs automatically when the action is finished.

An EnableInFalse scan would execute when the EnableIn parameter for the instruction is false. When an Add-On instruction is executed in the false condition and has an EnableInFalse routine defined and enabled, any required Parameters have their data passed. Values are passed to Input Parameters from their arguments in the instruction call. Values are passed out of Output Parameters to their arguments defined in the instruction call.

If the EnableInFalse routine has not been enabled, the only action performed for the Add-On instruction in the false condition is that the values are passed to any required Input Parameters in ladder logic.

Change History Tab

The Change History tab displays the latest edit information that has been tracked by the software (see Figure 5-7). The screen identifies who created and last edited the instruction based on the Windows user name at the time of the change.

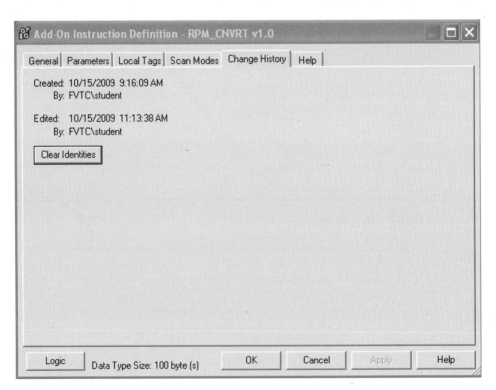

Figure 5-7   Change History screen.

Help Tab

The Help tab (see Figure 5-8) shows how the help file will appear and also allows you to edit the Help file. Much of the help file is automatically developed by the software as you develop the instruction. The software uses the Name, Revision, Descriptions, and Parameters definitions to build the help file. The developer of the Add-On instruction enters a longer and more thorough explanation of the instruction in the Extended Description Text box. The Instruction Help Preview shows how the Add-On instruction will appear in the various languages, based on Parameters defined as Required or Visible.

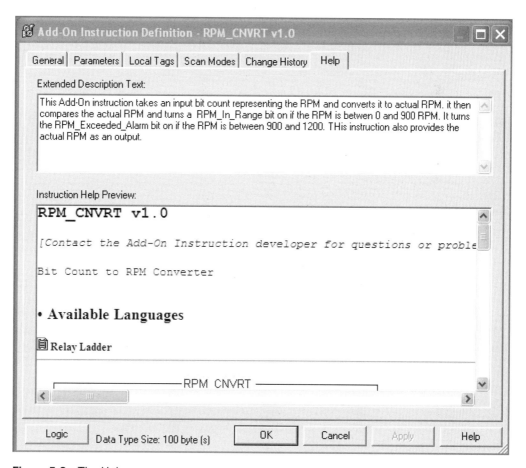

Figure 5-8   The Help screen.

Figure 5-9 shows the controller organizer after the information was entered for the new Add-On instruction.

Figure 5-9 The controller organizer showing the name of the new Add-On instruction.

PARAMETERS AND LOCAL TAG USAGE IN ADD-ON INSTRUCTIONS

Parameters and Local Tags are used to define the data type that is used when the instruction executes. The software builds the associated data type. Local Tags are added as hidden members.

Determining Whether Parameters Should Be Visible or Required

To ensure that data is passed into the Add-On instruction, you can use required Parameters. A required Parameter must be passed as an argument in order for a call to the instruction to pass verification when programming.

In Ladder Diagram and Structured Text programming, this is done by specifying an argument tag for these Parameters. In a Function Block Diagram, required input and output Parameters must be connected, and InOut Parameters must have an argument tag. If a required Parameter does not have an argument associated, then the routine containing the call to the Add-On instruction will pass verification. Making an output Parameter visible is useful even if you don't need to pass the Parameter value out to an argument, but you do want to display its value for troubleshooting.

Required Parameters are always visible, and InOut Parameters are always required and visible. All Input and Output Parameters, regardless of being marked as Required or Visible, can be accessed as a member of the instruction's tag.

Figure 5-10 shows the effects of Input, Output, and InOut visible settings.

LOGIC ROUTINE

The Logic routine of the Add-On instruction defines the primary functionality of the instruction. It is the code that executes whenever the instruction is called. Figure 5-11 shows the logic for the Add-On instruction named RPM_CNVRT. This logic takes a bit

Type	Required	Visible	Ladder Value	Ladder Argument	Function Block Must Connect	Function Block Argument	Function Block Diagram Change Visibility	Structured Text Argument
Input	Yes	Yes	Yes	Yes	Yes	N/A	No	Yes
Input	No	Yes	Yes	No	No	N/A	Yes	No
Input	No	No	No	No	No	N/A	Yes	No
Output	Yes	Yes	Yes	Yes	Yes	N/A	Yes	No
Output	No	Yes	Yes	No	No	N/A	No	Yes
Output	No	No	No	No	No	N/A	Yes	No
InOut	Yes	Yes	No	Yes	N/A	Yes	No	Yes

Figure 5-10 Input, Output, and InOut: required and visible settings effects.

count representing the speed of a motor from an analog input card and converts it to an actual RPM value though a scale (SCL) instruction. One limit (LIM) instruction sets an output bit named RPM_In_Range if the RPM is between 0 and 900. The second LIM instruction sets an output bit named RPM_Exceeded_Alarm if the RPM is between 900 and 1200. This Add-On instruction was developed in the Function Block language.

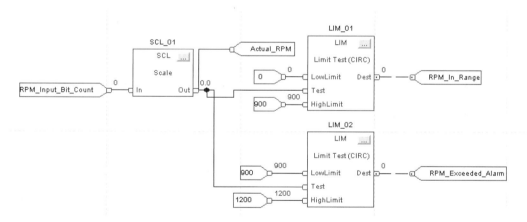

Figure 5-11 Logic for an Add-On instruction named RPM_CNVRT.

AVAILABLE LANGUAGES

Add-On instructions can be created in Ladder Diagram, Function Block Diagram, or Structured Text. Once the Add-On instruction has been created, it can be called from any of the RSLogix 5000 languages. An Add-On instruction written in one language can be used as an instruction through a call in another language. Instructions can also be stored so that they can be used in new projects.

Figure 5-12 shows how the instruction appears in a Function Block routine after it is developed. Note that it appears to be one Function Block instruction with one input and three outputs. The logic for the instruction had three Function Block instructions but the result is a simpler Add-On instruction.

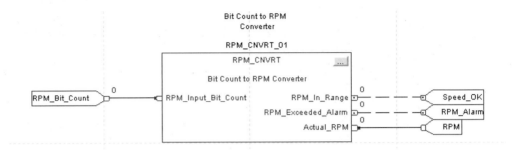

Figure 5-12 The completed Add-On instruction as it looks in a Function Block routine.

Figure 5-13 shows how the Add-On instruction appears when used in ladder logic in a routine.

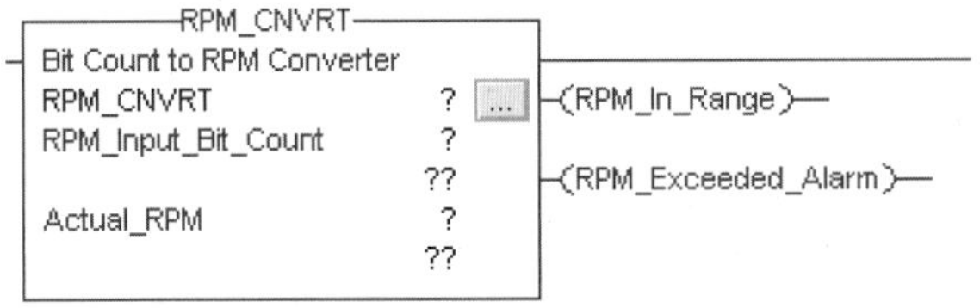

```
        ┌──────RPM_CNVRT──────────────┐──────────────────────
        ┤ Bit Count to RPM Converter  │
        │ RPM_CNVRT            ?  ┌───┐├─(RPM_In_Range)─
        │ RPM_Input_Bit_Count  ?  │...││
        │                     ??  └───┘├─(RPM_Exceeded_Alarm)─
        │ Actual_RPM           ?       │
        │                     ??       │
        └─────────────────────────────┘
```

Figure 5-13   Instruction as it appears as a ladder logic instruction.

Figure 5-14 shows how the Add-On instruction appears when used as a Structured Text instruction in a routine. Note the parameters in the parenthesis.

```
RPM_CNVRT(RPM_CNVRT,RPM_Input_Bit_Count,Actual_RPM);
```

Figure 5-14   How the Add-On instruction appears as a Structured Text instruction.

SOURCE PROTECTION

Defining Source Protection

You can protect the Add-On instructions that you develop. Source protection can be used to hide or protect proprietary code. You can apply source protection to protect your Add-On instructions or to prevent unintended edits. Source-protecting Add-On instructions can protect the Add-On instructions you develop and also prevent unwanted changes.

Applying Source Protection

Source protection capability can be used to limit a user of your Add-On instruction to read-only access or to allow no access to the internal logic or the Local Tags used by the instruction.

You can protect the use of your Add-On instructions and their modification with a source key file when you distribute them. This lets you stop unwanted changes to your instruction and protect your proprietary instructions. You may want to protect the source definition of an Add-On instruction from the view of a user. This may be due to the proprietary nature of the logic or for revision control. There are two types of protection that can be applied: Source Protected and Source Protected with Viewable Option.

Source Protected
Users without the source key cannot view the logic or Local Tags or make any edits to the Add-On instruction.

Source Protected with Viewable Option
Users without the source key can view all components of the Add-On instruction including its logic and Local Tags but are prevented from making any edits to the instruction.

Enabling the Source Protection Feature

If Source Protection is unavailable and not listed in your menus, you must enable the Source Protection feature. Source Protection can be enabled during the RSLogix 5000 system install, by enabling the feature. If this was not done, there is a tool on the installation CD named RS5KSrcPtc.exe.

If it is the first time the Source Protection has been configured, a dialog displays asking you to configure a source key file location. You must enter the location of (or browse to) the source key; then click **OK**.

Configuring Source Protection

To source protect an Add-On instruction, do the following:

1. In RSLogix 5000, select Tools → Security → Configure Source Protection to access the source configuration dialog.
2. Click **Specify** to identify the location for the sk.dat Source Key File. Click **OK** in both dialogs to accept. Click **Yes** to confirm creation of the sk.dat Key File.
3. Select the component you want to protect. Click **Protect**.
4. The Source Key Entry dialog displays. This is where the individual source key for this component is entered. The source keys follow the conventions for routine source keys. Next, expand the Add-On Instructions folder to view the instructions available to apply source protection. The components include all Routines and Add-On Instruction definitions in the project. Select the Add-On Instruction you wish to protect and click **Protect**. Enter a name for the source key for this component or use an existing one.

5. If you wish to allow users to be able to see the logic and Local Tags, check the box named Allow viewing of component(s). Before you click **Close**, you must first either click **Clear** or click **Disable Ability to Configure Source Protection**, or if you wish to observe the source protection settings, you must remove the sk.dat file from the PC so that the source key is no longer present.
6. Close and save the project.

QUESTIONS

1. Which languages can be used to develop Add-On instructions?
2. How is the documentation developed for an Add-On instruction?
3. How can an Add-On instruction be used to make logic less complex?
4. Can the logic in an Add-On instruction be made secure so that no one can see proprietary logic?
5. Utilize the following logic to develop an Add-On instruction. Stop, Start, and Run should all be visible and required Parameters in the instruction. Develop and test the instruction in ladder logic, structured text, and function block.

6. Utilize the following logic to develop an Add-On instruction. The parameters are shown below the logic. Develop the Add-On instruction and test it in ladder logic, structured text, and function block.

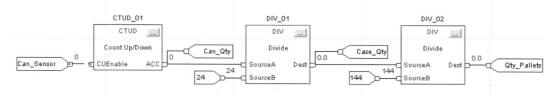

Name	Usage	Default	Force Mask	Style	Data Type	Description
+ Can_Qty	Output	0		Decimal	DINT	
Can_Sensor	Input	0		Decimal	BOOL	
+ Case_Qty	Output	0		Decimal	DINT	
+ CTUD_01	Local	{...}	{...}		FBD_COUNTER	
+ DIV_01	Local	{...}	{...}		FBD_MATH	
+ DIV_02	Local	{...}	{...}		FBD_MATH	
EnableIn	Input	1		Decimal	BOOL	Enable Input - Sys...
EnableOut	Output	0		Decimal	BOOL	Enable Output - S...
+ Qty_Pallets	Output	0		Decimal	DINT	

Scope: Case_Pack   Show...   Show All

Data Context: Case_Pack <definition>

APPENDIX

A

Starting a New Project

There are three basic steps in starting a new project. First you name the project and configure the project for the correct CPU and slot, software revision, and chassis type. Next you should set a path to the CPU. Third you add the required I/O modules. RSLinx should be configured with the correct Communications Driver to be able to communicate with the CLX controller you would like to program.

Open RSLOGIX 5000 and Select File and then New Project. The screen shown in Figure A-1 should appear. First you must choose the correct type of processor. In this

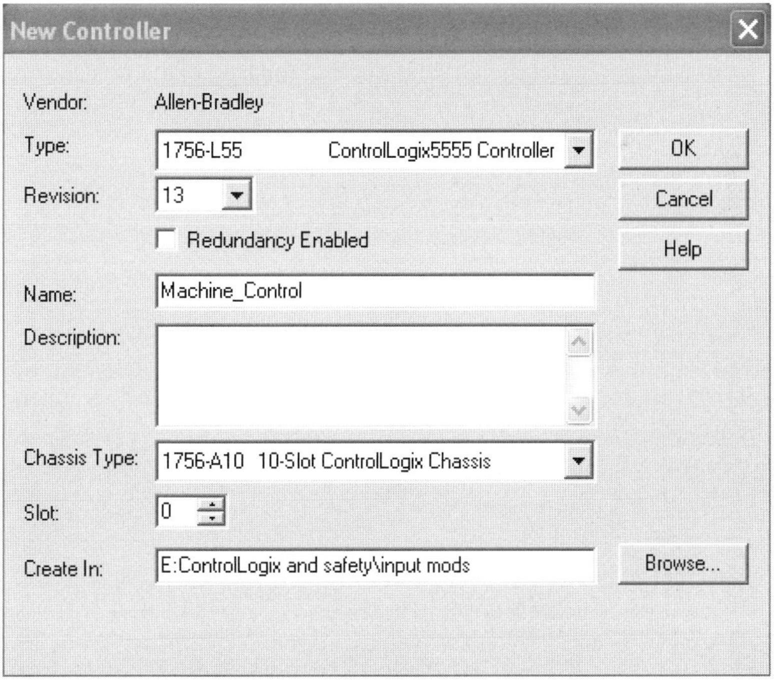

Figure A-1 New project configuration screen.

example the processor is a 1756-L55. Next choose the correct RSLogix 5000 Revision (software level). It is 13 for this example. Next you must name the project. Machine_Control was the name chosen in this example. You must also be sure to choose the correct chassis type. You may change the location where it will be created. In this example the path chosen was E:\ControlLogix and safety\input mods.

A path to the controller's CPU should be set next (see Figure A-2). In this figure the path is blank yet. You can choose the down arrow of the RSWho icon to the right to choose a path. If you select the RSWho icon, the screen in Figure A-3 will appear.

Figure A-2 Project screen. Note that the path has not been set yet.

Click on the CPU you would like to program and then choose Set Project Path. Figure A-3 shows that the CPU in slot 0 in the backplane was chosen for the CPU. In this example the path went through an RSLINX Ethernet devices driver.

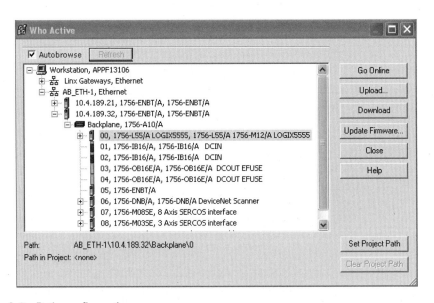

Figure A-3 Path configuration screen.

Then you can select Close to close the RSWho screen and you will see that the path now appears as shown in Figure A-4.

Figure A-4   Note that the path has been set.

Next you should add any modules that the application will require. Figure A-5 shows the Controller Organizer. Note that I/O Configuration is shown on the bottom but no modules have been added yet.

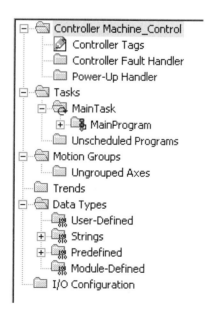

Figure A-5   Controller Organizer with no modules added yet.

To add a module, you must right-click on the I/O Configuration icon and choose New Module. The screen shown in Figure A-6 will then appear. You can choose the module from the list of available modules. In this example a 1756-IB16 was chosen for the input module. Select OK, and the screen in Figure A-7 will appear.

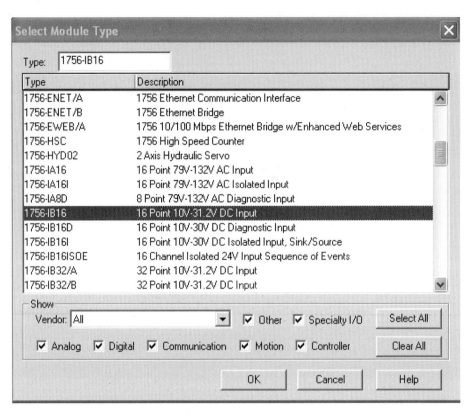

Figure A-6 List of modules that can be added. Note that a 1756-IB16 input module is high-lighted in the list.

Next you must choose the Major Revision for the module you are adding. In this example the Major Revision was 2. The revision level is shown on the side of the module. You can also use RSLinx to find the revision level of the module. The use of RSLinx will assure that you have the right revision in case the module firmware was upgraded.

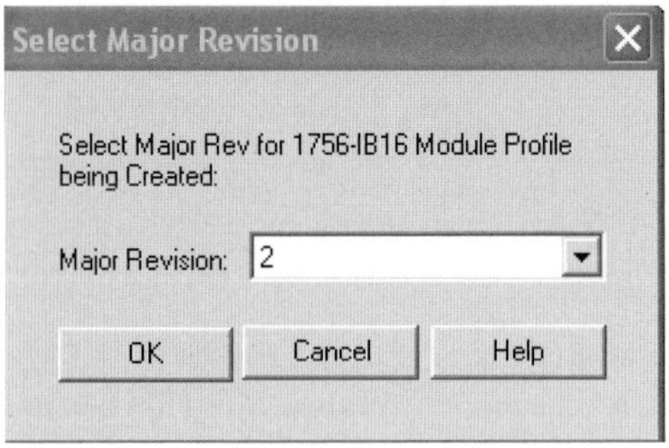

Figure A-7 Select Major Revision screen.

Figure A-8 shows the Module Properties screen for the input module that is being added. The name entered was Input_Mod_1. It is located in slot 1. The keying method chosen was Compatible Module.

Module Properties - Local:1 (1756-IB16 2.1)   ✕

Type:	1756-IB16 16 Point 10V-31.2V DC Input
Vendor:	Allen-Bradley
Parent:	Local
Name:	Input_Module_1 Slot: 1
Description:	
Comm Format:	Input Data
Revision:	2 1 Electronic Keying: Compatible Module

Cancel   < Back   Next >   Finish >>   Help

Figure A-8   Module Properties screen for the input module.

Figure A-9 shows the Controller Organizer after the input module was added.

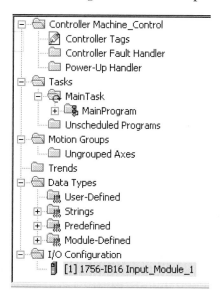

Figure A-9   Controller Organizer with input module added.

Next the output module will be added. Right click the I/O Configuration icon in the Controller Organizer and choose New Module. The screen shown in Figure A-10 shows the module selection screen that will appear. The output module for this example is a 1756-OB16E output module. This is a 16-output electronically fused module. Then select OK.

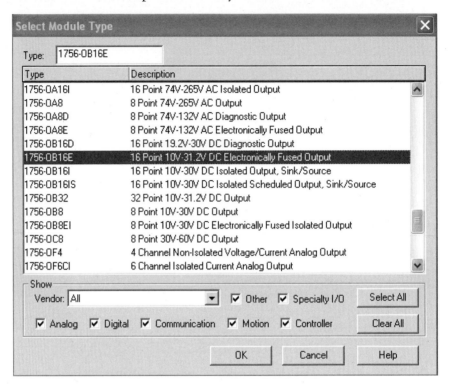

Figure A-10   Note that a 1756-OB16E was chosen in this example.

Next the Select Major Revision screen will appear (see Figure A-11). The Major Revision level for this module was 2. Select OK.

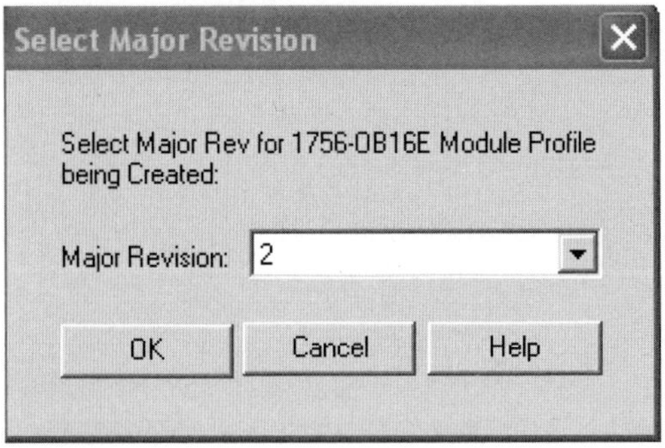

Figure A-11   Select Major Revision screen.

The Module Properties screen then appears (see Figure A-12). Enter a name for the module, choose the slot where it is located, enter the minor Revision level, and choose the keying method for this module. Then select Finish.

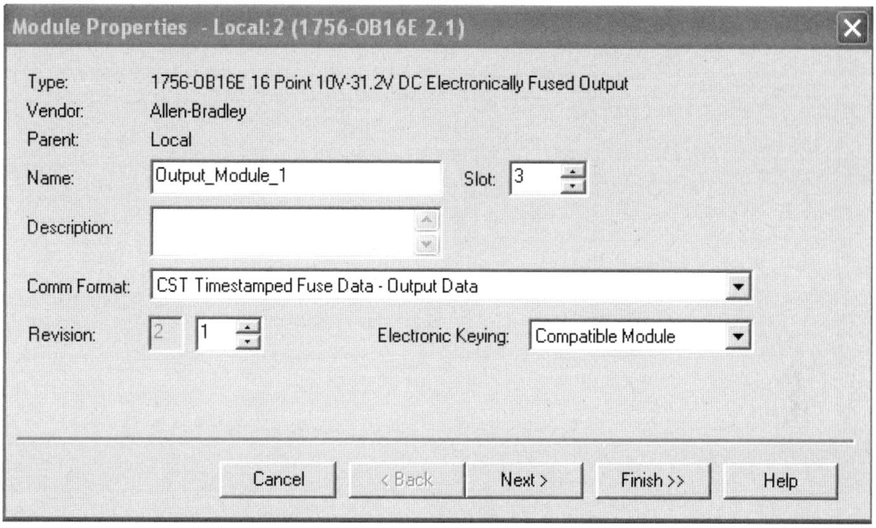

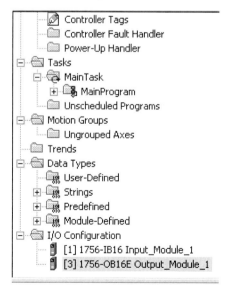

Figure A-12   Module Properties screen.

Note that this module was named Output_Module_1 in this example. The module is in slot 2. Compatible Module was chosen for the method of keying.

Figure A-13 shows the Controller Organizer after the modules have been added.

Figure A-13   Controller Organizer after an input and an output module were added.

You are now ready to program. Click on the + next to the MainProgram folder in the Controller Organizer shown in Figure A-13. Program Tags and MainRoutine will appear in the Controller Organizer as shown in Figure A-14. Double-click on MainRoutine, and the programming screen will open up as shown in Figure A-15. You are now ready to program.

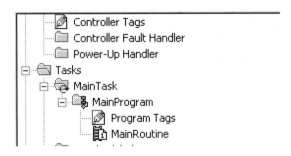

Figure A-14 MainRoutine in Controller Organizer.

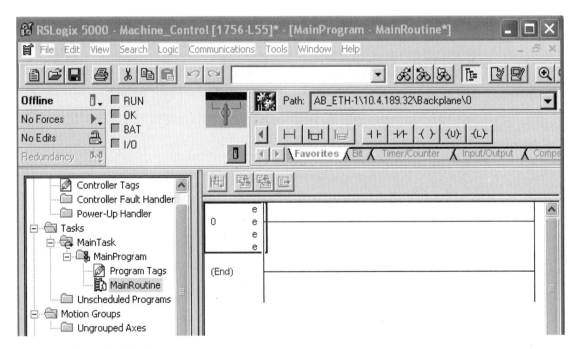

Figure A-15 Programming screen.

APPENDIX

B

Remote Inputs and Outputs

A controller in one chassis can own modules in another chassis. Remember that output modules can only have one owner but multiple controllers can own the same input module as long as they are configured exactly the same.

Figure B-1 shows an example of a simple system. There are two CLX systems in the figure. This appendix will show how to configure a project so that the controller in one chassis will own and control an output module in a different chassis. The project will be configured so that the controller on the left of Figure B-1 will own the output module in slot 3 of the CLX system on the right. Remember that no other controller can own the module. Note also that there is an Ethernet Bridge module in each chassis. ControlNet would also work, but this example uses Ethernet.

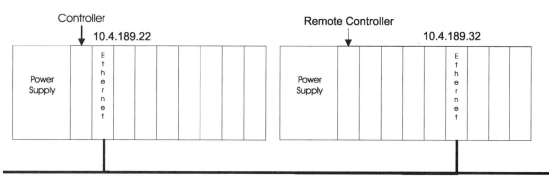

Figure B-1   Two CLX systems. Note that each chassis has an Ethernet module.

It is a relatively straightforward process to configure a project to use remote I/O. It really just involves adding cards that will provide a path under the I/O configuration for the project. Consider the system in Figure B-1. The path from the controller would be through the backplane of the Ethernet module that resides in the same chassis, then out the Ethernet module to the Ethernet module in the remote chassis, and then through the backplane to the module in slot 3. The path involves three modules and begins with the Ethernet module in the controller's chassis.

Figure B-2 shows the Project Organizer screen. You need to add modules under the I/O Configuration folder. To do this, you right-click on I/O Configuration and choose New Module. The screen shown in Figure B-3 will then appear.

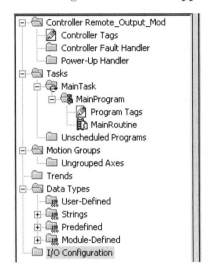

Figure B-2 Controller Organizer before any modules were added.

Select the communications module. In this example it is a 1756-EBET/B module. Then click on OK.

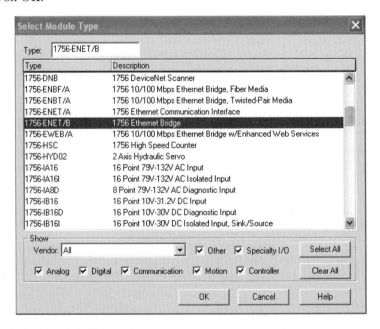

Figure B-3 Select Module Type screen.

Next the module must be configured (see Figure B-4). The IP address for this module is 10.4.189.22. The module is in slot 1 of the chassis so 1 was entered for Slot. The correct Revision level should be chosen for the module. A name is also entered for the module. In this example the module was named Ethernet_Local. Compatible Module was chosen for the Electronic Keying type. Then Finish can be selected, and the screen in Figure B-5 will appear.

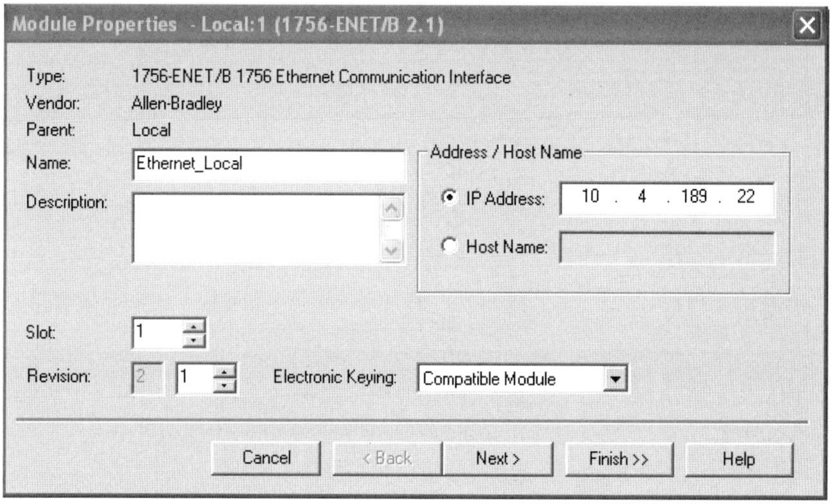

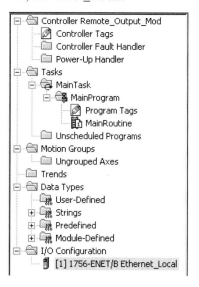

Figure B-4   Module Properties screen.

Figure B-5 shows the Controller Organizer after the Ethernet module was added. Notice the name of the module, Ethernet_Local.

Figure B-5   Controller Organizer after the Ethernet module was added.

Next you must add the remote Ethernet module. This must be added under the Local Ethernet module that was just added. Right-click on the Ethernet module (Ethernet_Local) that was just added and choose New Module. The screen in Figure B-6 will appear. Choose the correct module. In this example the 1756-ENET/B module was chosen. Choose OK and the screen in Figure B-7 will appear.

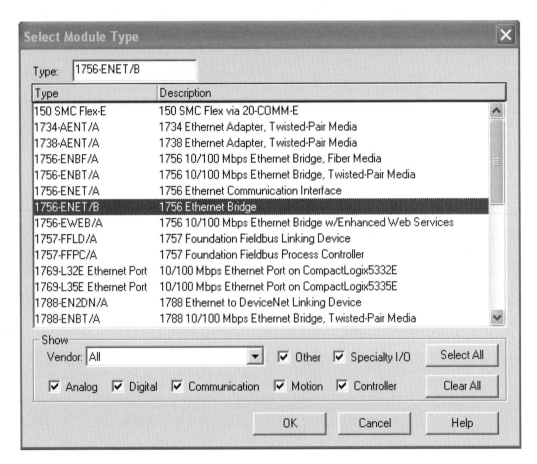

Figure B-6 Select Module Type screen.

Next the module's properties must be configured (see Figure B-7). The name Ethernet_Remote was entered for the name. The IP address was entered (10.4.189.32). The module is in slot 5, so 5 was entered for Slot. Make sure the correct Revision level is set. Compatible Module was chosen for the Electronic Keying type. Select Finish, and the screen in Figure B-8 will appear.

Module Properties - Ethernet_Local (1756-ENET/B 2.1)   ☒

Type: 1756-ENET/B 1756 Ethernet Communication Interface
Vendor: Allen-Bradley
Parent: Ethernet_Local
Name: [Ethernet_Remote]

Description: [   ]

┌─ Address / Host Name ─────────────────┐
│ ⊙ IP Address: [10 . 4 . 189 . 32] │
│   │
│ ○ Host Name: [   ] │
└──┘

Comm Format: [Rack Optimization   ▼]
Slot: [5 ⬍]   Chassis Size: [10 ⬍]
Revision: [2] [1 ⬍]   Electronic Keying: [Compatible Module ▼]

[Cancel]   [< Back]   [Next >]   [Finish >>]   [Help]

Figure B-7   Module Properties screen.

Figure B-8 shows the Controller Organizer after the remote Ethernet module was added. Note the name on the module is Ethernet_Remote. Also note that it was added under the first Ethernet module.

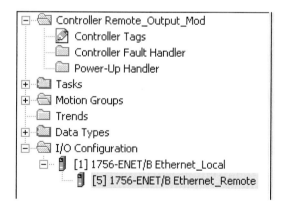

```
⊟ 🗀 Controller Remote_Output_Mod
      🗎 Controller Tags
      🗀 Controller Fault Handler
      🗀 Power-Up Handler
⊞ 🗀 Tasks
⊞ 🗀 Motion Groups
   🗀 Trends
⊞ 🗀 Data Types
⊟ 🗀 I/O Configuration
   ⊟ 🔲 [1] 1756-ENET/B Ethernet_Local
         🔲 [5] 1756-ENET/B Ethernet_Remote
```

Figure B-8   Controller Organizer after the remote Ethernet module was added.

When this module is added, controller tags are automatically generated as shown in Figure B-9. They can be seen in the tag editor.

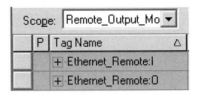

Figure B-9   Controller tags for the Ethernet module.

Next you need to add the actual I/O module, in this example an output module. Right-click on the second Ethernet module that was added. Choose New Module and the screen shown in Figure B-10 will appear. In this example a 1756-OB16E output module was chosen. Then select OK.

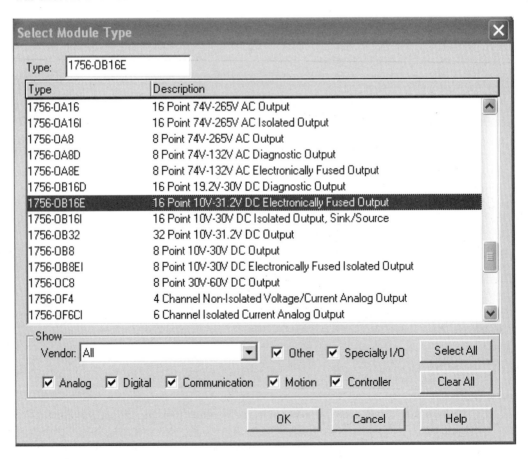

Figure B-10   Select Module Type screen.

The correct Major Revision level is set next (see Figure B-11). Choose OK, and the screen shown in Figure B-12 will appear.

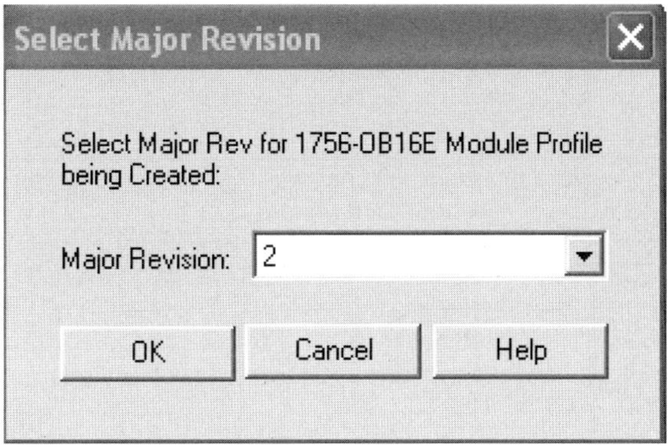

Figure B-11   Major Revision level entry screen.

In Figure B-12 the module was named Remote_Output_Module. The module resides in slot 3. Compatible Module was chosen for the Electronic Keying type. A Comm (communications) Format must be chosen. In this example Rack Optimization was chosen. Rack Optimization and the other choices are covered in Figure B-13. Choose Finish, and the screen in Figure B-14 will appear.

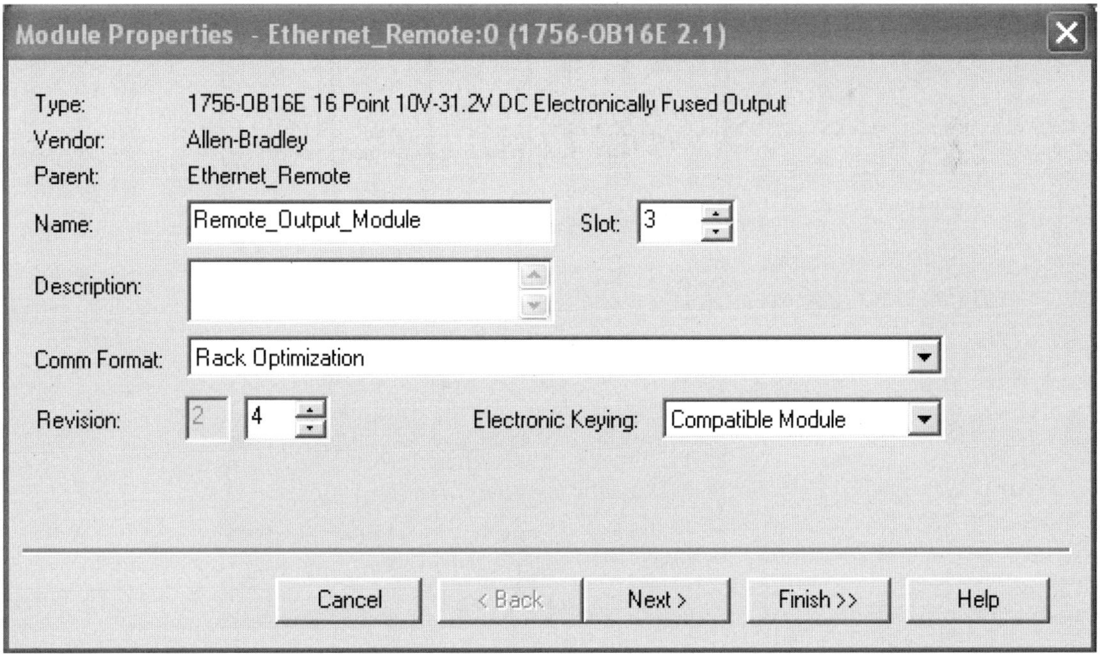

Figure B-12   Module Properties screen.

If	Select
The remote chassis contains only analog modules, diagnostic digital modules, fused output modules, or communication modules.	None
The remote chassis only contains standard, digital input and output modules (no diagnostic or fused output modules).	Rack Optimization
You want to receive I/O module and chassis slot information from a rack-optimized remote chassis owned by another controller	Listen-Only Rack Optimization

Figure B-13 Comm Format choices.

Figure B-14 shows the Controller Organizer after the output module was added. Notice the name on the module (Remote_Output_Module). The configuration is complete at this point. If you look in the tag editor screen you will see that tags have been automatically added for the module. The actual outputs are in the tag named Ethernet_ Remote:3:O (see Figure B-15).

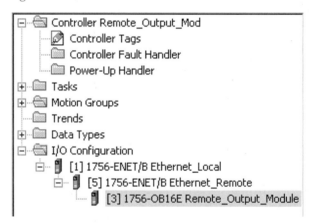

Figure B-14 Controller Organizer after the output module was added.

Figure B-15 Tag editor screen showing the tags that were added when the module was added.

APPENDIX

C

Producer Consumer

Producer/consumer tags provide an easy means for a processor to provide data to other controllers. By using the producer/consumer model, data can be transferred between processors without any logic. The user can choose the Requested Packet Interval (RPI) for the rate at which data should be updated. Data can be transferred between DINT-type tags, an array, or a User-Defined data type.

THE PRODUCED TAG

The produced tag is the easiest part to configure. In the CLX controller that will be the producer, the user simply creates a controller Scope tag and configures it as produced. The name of the produced tag in this example is Produced_Tag (see Figure C-1). It is a DINT. Note the check mark in the P column. P stands for Produced.

Marking a tag produced enables the tag to be available to consumed tags in other controllers.

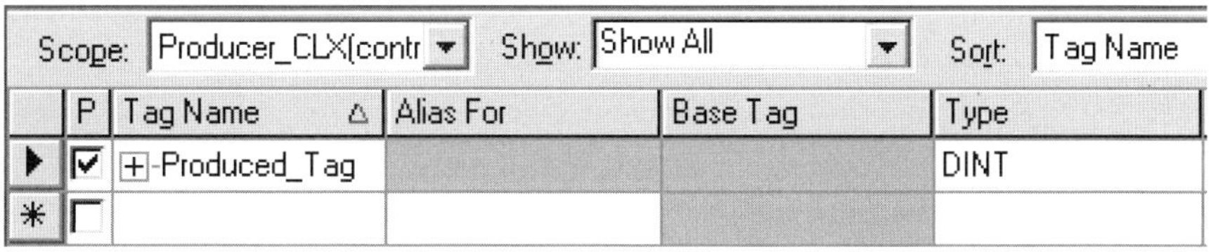

	P	Tag Name △	Alias For	Base Tag	Type
▶	✓	⊞-Produced_Tag			DINT
✱	☐				

Scope: Producer_CLX(contr ▼   Show: Show All ▼   Sort: Tag Name

Figure C-1 A controller Scope tag named Produced_Tag. Note the check mark in the P column. This makes the tag a produced tag.

Figure C-2 shows how it appears in the Tag Properties screen.

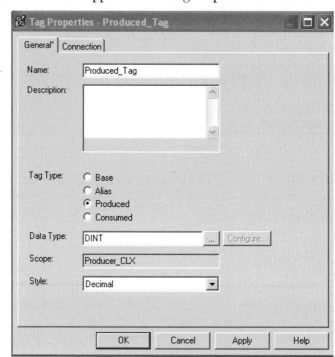

Figure C-2 Tag Properties screen for the tag. Note that the Tag Type is Produced. Note also the Connection tab on top of the screen.

If you click on the Connection tab, you can set the maximum number of consumers that are allowed to connect to the produced tag (see Figure C-3). The allowed values are 1 to 256. One other controller will be allowed access to this tag in this example. You can also send event triggers to consumers using the IOT instruction in logic.

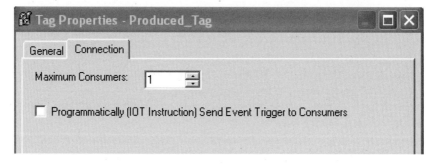

Figure C-3 Tag Properties, Connection parameters. In this example only one other controller (consumer) will be allowed to get the data for this tag because 1 was entered in Maximum Consumers.

That is all you have to do in the Producer controller.

THE CONSUMED TAG

Next a project must be developed for the controller who will be the consumer of tag data from the produced tag. Create a new project in a different controller. Set a path and add any I/C modules you need for your application. In this example only communication modules will be required.

First, you must create a communications path to the processor where the tag is being produced. Then the consumed tag can be created.

Figure C-4 shows an illustration of the two CLX systems. Note that each chassis has a controller and an Ethernet module. They are both attached to the Ethernet network. Information for this example will use the Ethernet communications network. Other networks such as ControlNet could have been used. Note that the Ethernet module is in slot 7 in the Producer Controller and in slot 6 in the Consumer Controller.

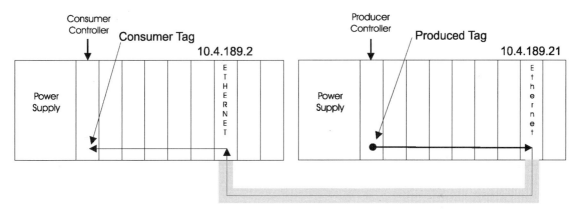

Figure C-4   The CLX systems used in this example.

In the CLX project organizer (see Figure C-5), a path will need to be configured to the processor that has the produced tag. Referring back to Figure C-4, you see the module path from the consumer CLX must be established to the producer CLX. The path will be through the Ethernet module in the consumer CLX chassis, through the Ethernet module in the producer chassis, and finally to the controller (CPU) in the producer chassis.

To add the Ethernet module in the consumer chassis, right-click on the I/O Configuration folder in the Controller Organizer (see Figure C-5). After you right-click, you will choose New Module and the screen shown in Figure C-6 will appear.

Figure C-5   Controller Organizer before the modules have been added to create a path to the Produced tag.

Next you must choose the correct module. The module in the consumer chassis is a 1756-ENET/B module. This is a 1756 Ethernet Bridge module. Then select OK. The screen shown in Figure C-7 will appear so that you can configure the module.

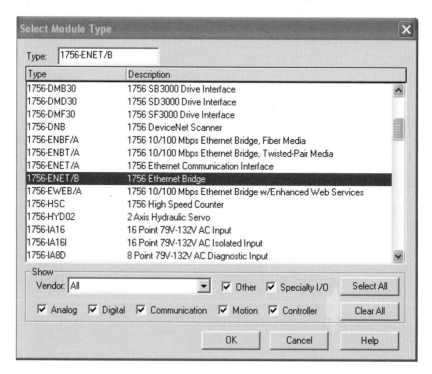

Figure C-6   The screen to select the module type.

In the Module Properties screen in Figure C-7 you enter a name for the module. You must enter the IP address of the Ethernet module. You also enter the Slot that the module resides in and the Revision. You can choose the keying method for the module.

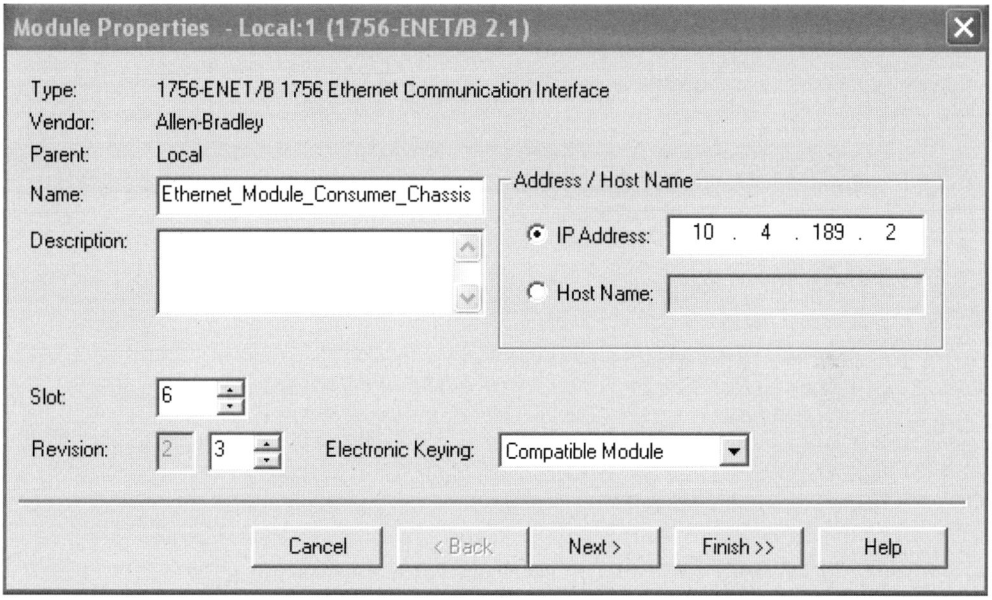

Figure C-7 Module Properties screen.

When you choose Finish, the Controller Organizer displays the module as shown in Figure C-8.

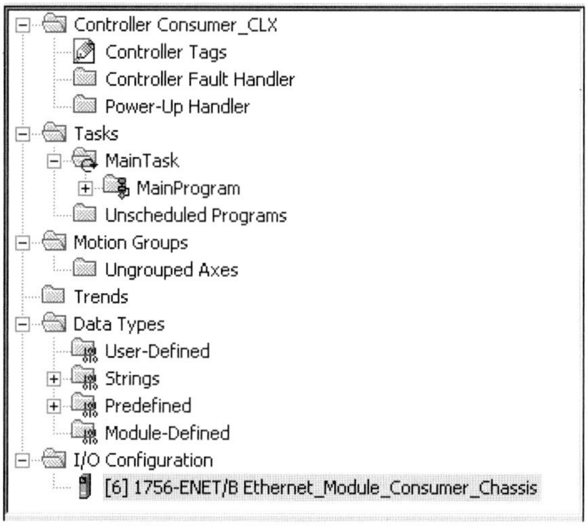

Figure C-8 Controller Organizer after the module was added.

Next you must add the Ethernet module in the other Chassis (producer CLX chassis). This one will be added under the first Ethernet module. Right-click on the Ethernet module that you just added. Choose New Module.

Next you will choose the correct module as shown in Figure C-9. The module in this example is a 1756-ENET/B module. This is a 1756 Ethernet Bridge module. Then select OK. The screen shown in Figure C-10 will appear.

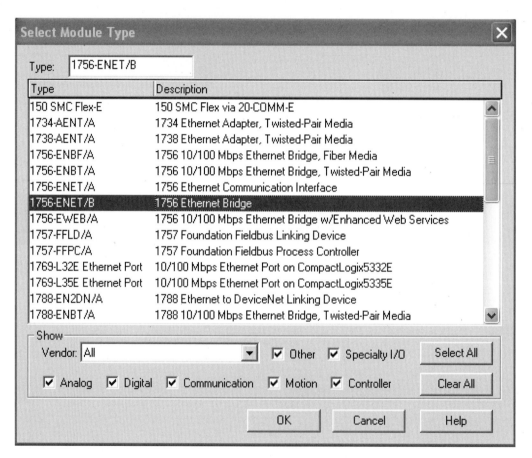

Figure C-9   Select Module Type screen.

In the Module Properties screen in Figure C-10, you enter a name for the module. You must enter the IP address of the Ethernet module. You also enter the Slot that the module resides in and the Revision level. You can choose the Electronic Keying method for the module. The last thing to choose is the Comm Format. Rack Optimization was chosen in this example.

Module Properties - Ethernet_Module_Consumer_Chassis (1756-ENET/B 2.1) ☒

Type: 1756-ENET/B 1756 Ethernet Communication Interface
Vendor: Allen-Bradley
Parent: Ethernet_Module_Consumer_Chassis
Name: Ethernet_Module_Produced_Chassis

Address / Host Name
◉ IP Address: 10 . 4 . 189 . 21
○ Host Name:

Description:

Comm Format: Rack Optimization ▼
Slot: 7 ⏷ Chassis Size: 10 ⏷
Revision: 2 3 ⏷ Electronic Keying: Compatible Module ▼

Cancel | < Back | Next > | Finish >> | Help

Figure C-10 Module Properties screen.

Figure C-11 shows the Project Organizer screen after the Ethernet module in the producer chassis was added.

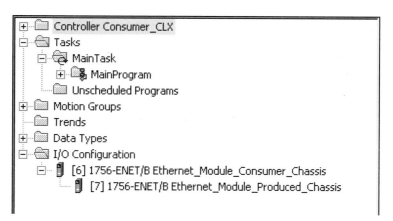

⊞ ▢ Controller Consumer_CLX
⊟ ▢ Tasks
   ⊟ ▢ MainTask
      ⊞ ▢ MainProgram
      ▢ Unscheduled Programs
⊞ ▢ Motion Groups
  ▢ Trends
⊞ ▢ Data Types
⊟ ▢ I/O Configuration
   ⊟ ▮ [6] 1756-ENET/B Ethernet_Module_Consumer_Chassis
      ▮ [7] 1756-ENET/B Ethernet_Module_Produced_Chassis

Figure C-11 Project Organizer screen after the Ethernet module in the producer chassis was added.

So far you have a path from the Ethernet module in the consumer CLX chassis to the Ethernet module in the producer chassis. Lastly you must add the controller (CPU) that produces the tag. Right-click on the Ethernet module you just added and choose New Module. The screen in Figure C-12 will appear. Choose the type of controller, 1756-L55 in this example. Then click OK, and the screen shown in Figure C-13 will appear.

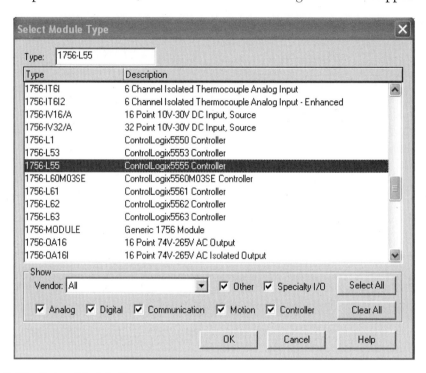

Figure C-12 Select Module Type screen.

Choose the correct Major Revision level for the producer controller.

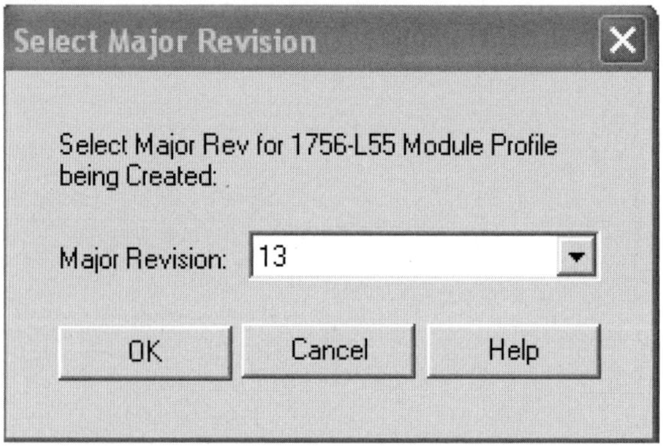

Figure C-13 Select Major Revision level screen.

After you enter the Major Revision level, click OK, and the screen shown in Figure C-14 will appear. Enter a name for the producer controller as well as the Slot that it resides in. Then click the Finish button.

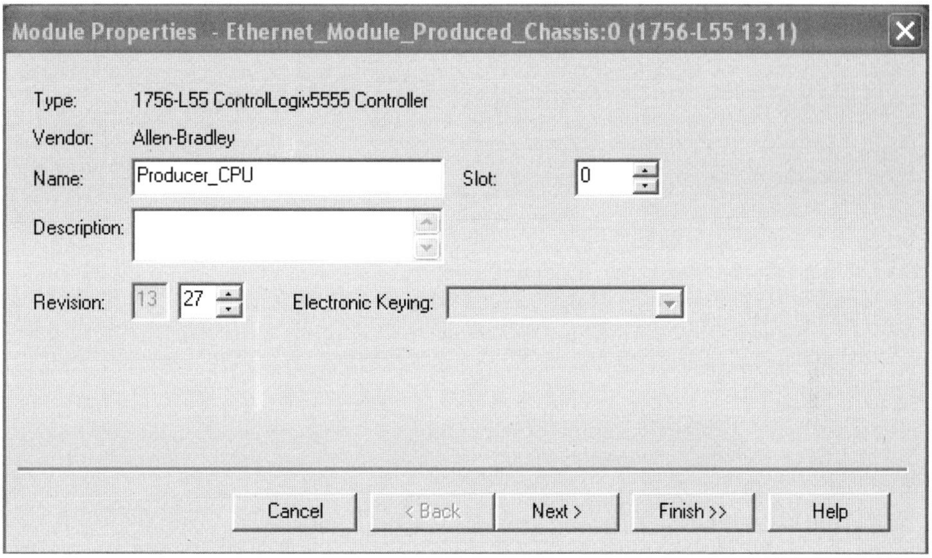

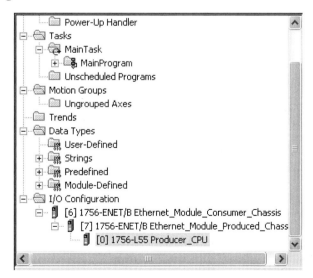

Figure C-14   Module Properties screen.

Now the Project Organizer screen looks like Figure C-15. Note that this now provides the path to get the data.

Figure C-15   Project Organizer screen after the producer controller module was added.

The last thing to do is create the consumed tag. In this example the consumed tag was named Consumed_Tag (see Figure C-16). It is a DINT type for this example.

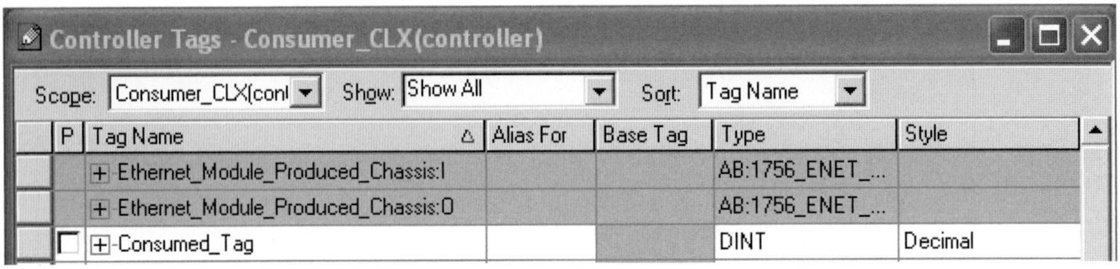

Figure C-16 Tag editor showing the consumed tag that is named Consumed_Tag.

Figure C-17 shows the Tag Properties screen. Note the Tag Type is Consumed and the Data Type DINT.

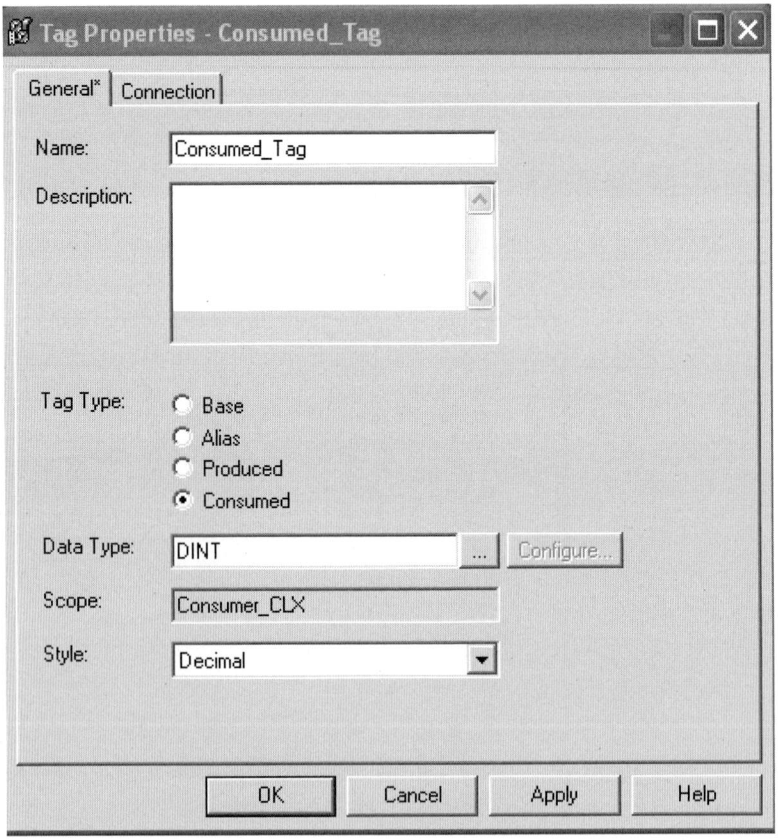

Figure C-17 Tag Properties screen.

Next click on the Connection tab on the Tag Properties screen. The screen shown in Figure C-18 appears. In this screen the Producer controller is chosen from the drop-down list (Producer_CPU in this example). Then the name of the produced tag in the remote controller is entered (Produced_Tag in this example). A value is also entered for the RPI (5 in this example). This determines how often the tag is updated.

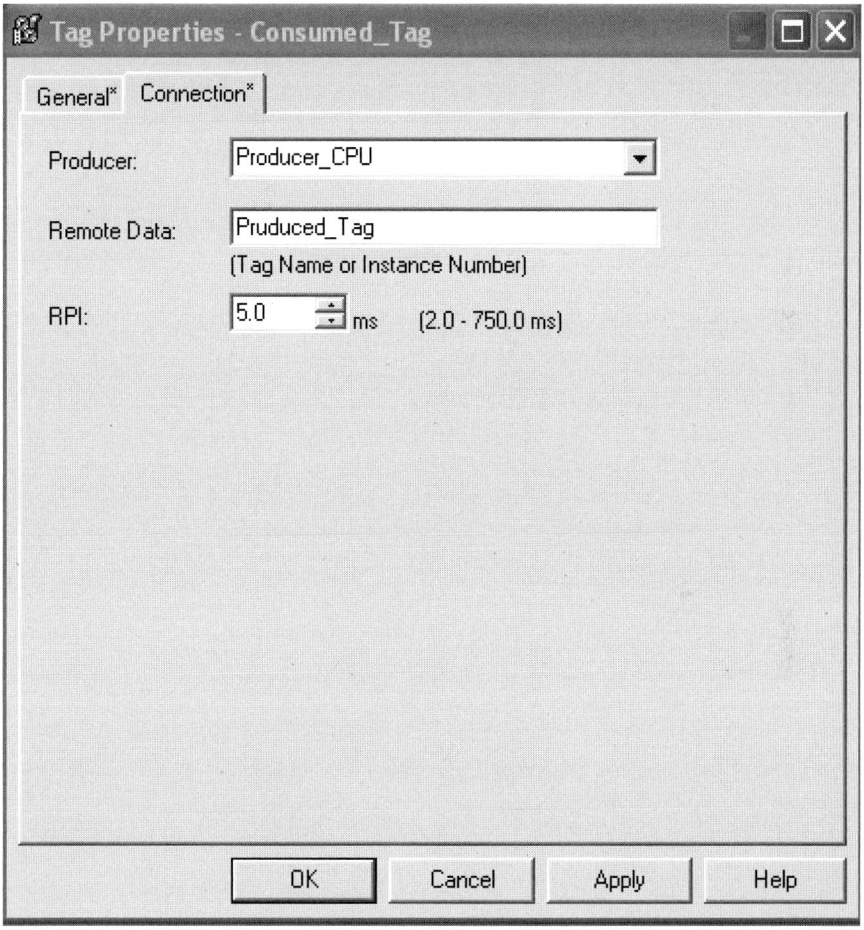

Figure C-18   Tag Properties screen.

At this point both controllers should be put into run mode. Type a number into the produced tag. Figure C-19 shows the number 55 has been entered into the Produced_Tag in the tag editor. The number should then appear in the consumed tag in the other controller. Figure C-20 shows the tag editor screen in monitor mode.

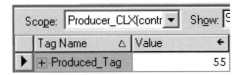

Figure C-19 Monitor mode in the tag editor screen in the producer controller.

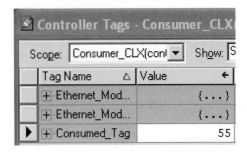

Figure C-20 Monitor mode in the tag editor screen in the consumer controller.

APPENDIX

D

Messaging

This appendix will show how messaging can be configured and accomplished between two CLX controllers.

The producer/consumer model is very efficient for transferring data between processors, but if the data transfer does not need to occur at periodic intervals, you may be able to conserve network bandwidth using the message (MSG) instruction. Using the MSG instruction, data can even be received (or sent) from another processor, even if that processor is not present in the I/O Configuration tree.

For this example, consider an MSG instruction that will read data from another ControlLogix processor and store that data in a memory location in your controller.

Below, you can see the path to take to connect to the target processor. Once the connection is made, the Temp_In array tag in the controller on the left will receive data from the Temp_Remote array tag in the controller on the right each time the MSG instruction is executed.

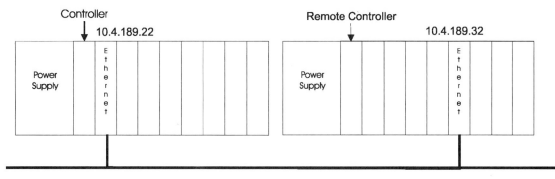

Figure D-1 The two CLX systems.

CREATE REMOTE TAGS

The first step is to create the tag in the remote CLX controller. In this example the tag is in the controller on the left of Figure D-1. Figures D-2 and D-3 show a DINT-type array tag named Temp_Remote was created and has ten members. It was created as a controller Scope tag. The Temp_Remote[0] array will be the memory location that another controller can read from. Data will be put in this tag's members, so the controller that reads from the tag can test the connection.

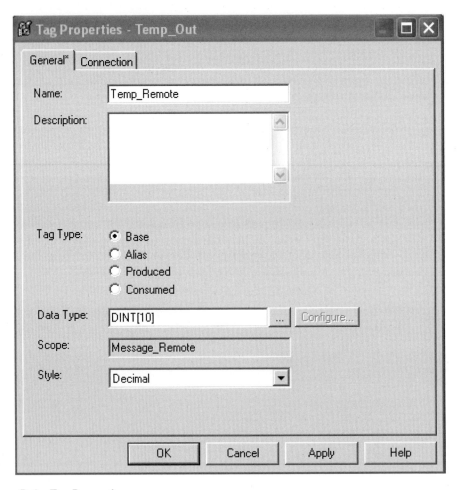

Figure D-2   Tag Properties screen.

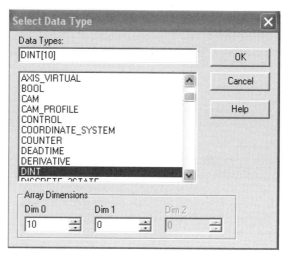

Figure D-3 Select Data Type screen.

SETTING UP THE CLX CONTROLLER FOR MESSAGING

Next you need to configure the CLX controller that will be utilizing the MSG command to get information. You need to establish a communications path for it to access the information from the remote CLX controller. You will utilize the Ethernet modules in the two controller chassis for the communications path. Figure D-4 shows the Controller Organizer before any modules have been added.

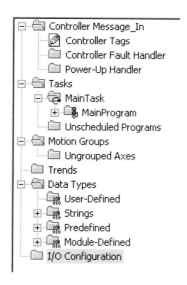

Figure D-4 Controller Organizer screen before any modules have been added.

The first module that was added in this example was the Ethernet module. To do this, right-click on the I/O Configuration shown in Figure D-4. Choose New Module. The screen shown in Figure D-5 appears. A 1756-ENET/B module is chosen. Click on OK, and the screen in Figure D-6 appears.

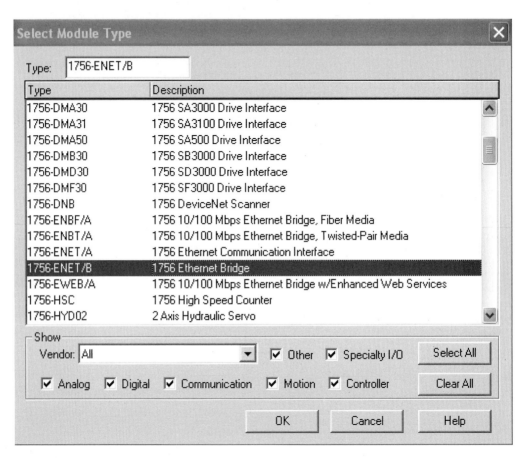

Figure D-5 Select Module Type screen.

In the screen shown in Figure D-6 the IP address was entered (10.4.189.22). The Ethernet module is located in slot 1 so 1 was entered in the Slot. Compatible Module was chosen for the Electronic Keying method. Click on the Finish button, and the screen in Figure D-7 appears.

Module Properties - Local:1 (1756-ENET/B 2.1)   ☒

Type:	1756-ENET/B 1756 Ethernet Communication Interface
Vendor:	Allen-Bradley
Parent:	Local
Name:	Ethernet_Local
Description:	

Address / Host Name

     ⦿ IP Address: 10 . 4 . 189 . 22

     ⦾ Host Name:

Slot: 1

Revision: 2 1   Electronic Keying: Compatible Module ▼

Cancel   < Back   Next >   Finish >>   Help

Figure D-6 Module Properties screen.

Figure D-7 shows the Controller Organizer after the local Ethernet module was added.

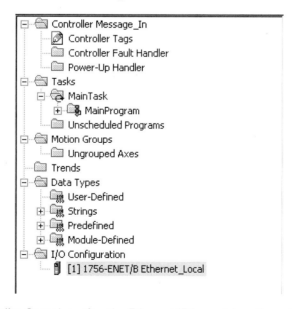

Figure D-7   Controller Organizer after the Ethernet (Ethernet_Local) module was added.

Next you must add the Ethernet module in the remote chassis. Right-click on the Ethernet module that was just added and choose New Module. Choose the correct module from the list shown in Figure D-8. In this example a 1756-ENET/B module was chosen.

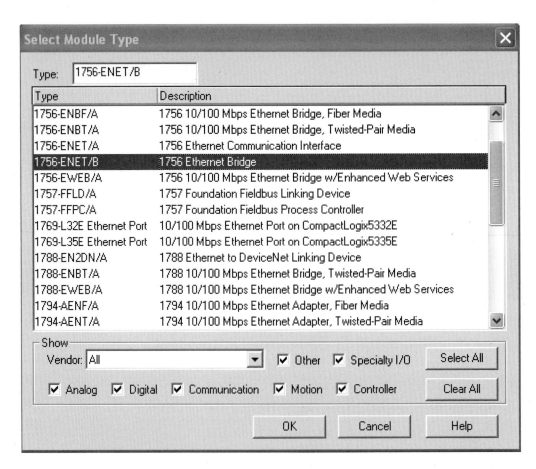

Figure D-8 Select Module Type screen.

The IP address for the remote Ethernet module was entered (10.4.189.32) in this example (see Figure D-9). A 5 was entered for the Slot, and Compatible Module was chosen for the Electronic Keying method.

Module Properties - Ethernet_Local (1756-ENET/B 2.1)

Type:	1756-ENET/B 1756 Ethernet Communication Interface
Vendor:	Allen-Bradley
Parent:	Ethernet_Local

Name: Ethernet_Remote

Description:

Address / Host Name

- ⦿ IP Address: 10 . 4 . 189 . 32
- ○ Host Name:

Comm Format: Rack Optimization

Slot: 5   Chassis Size: 10

Revision: 2 1   Electronic Keying: Compatible Module

Cancel   < Back   Next >   Finish >>   Help

Figure D-9   Module Properties screen.

Figure D-10 shows the Controller Organizer after the second Ethernet module was added. Notice it was added under the first Ethernet module.

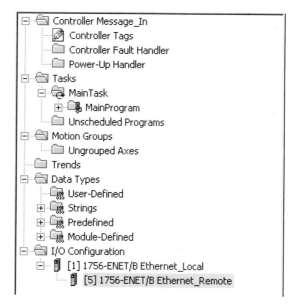

```
Controller Message_In
    Controller Tags
    Controller Fault Handler
    Power-Up Handler
Tasks
    MainTask
        MainProgram
        Unscheduled Programs
Motion Groups
    Ungrouped Axes
Trends
Data Types
    User-Defined
    Strings
    Predefined
    Module-Defined
I/O Configuration
    [1] 1756-ENET/B Ethernet_Local
        [5] 1756-ENET/B Ethernet_Remote
```

Figure D-10   Controller Organizer screen after the Ethernet (Ethernet_Remote) module was added.

The last thing that needs to be done is to add the CLX controller module that is located in the remote chassis. Right-click on the Ethernet module that was just added (see Figure D-10). Choose New Module. Choose the correct controller from the list shown in Figure D-11. In this example 1756-L55 was chosen. Then select OK.

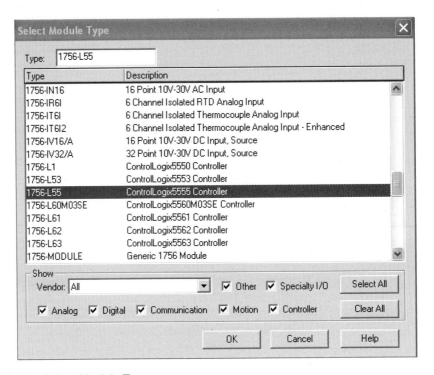

Figure D-11   Select Module Type screen.

Enter the Major Revision as shown in Figure D-12, and then choose OK.

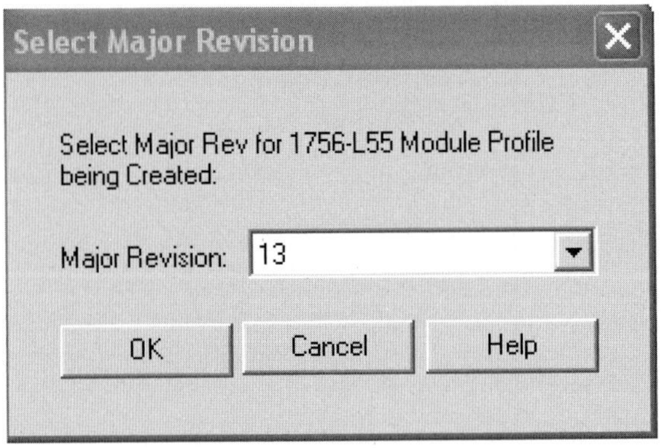

Figure D-12   Select Major Revision level screen.

Next enter the name for the module as shown in Figure D-13. Choose the correct Slot, 0 in this example. Then choose Finish.

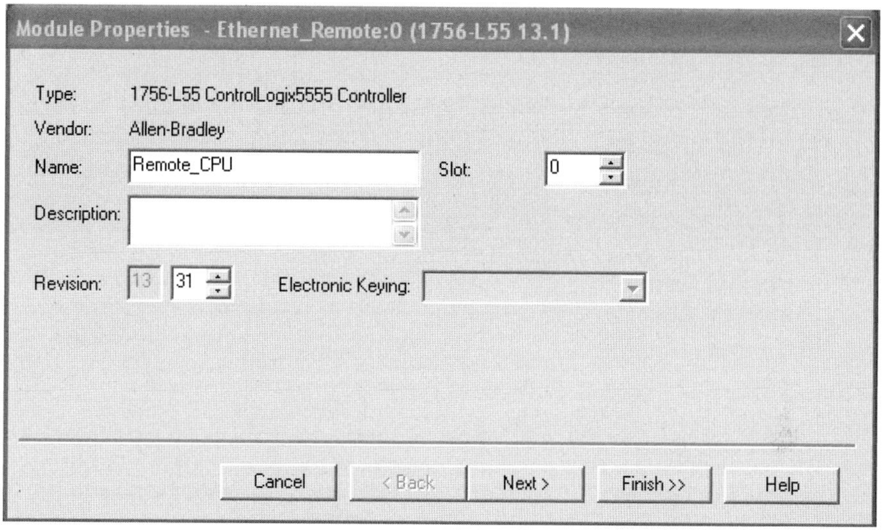

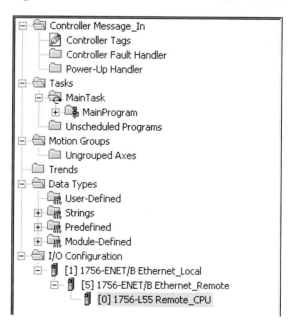

Figure D-13 Module Properties screen.

The Controller Organizer now shows the controller has been added (see Figure D-14).

Figure D-14 Controller Organizer screen after the controller (Remote_CPU) module was added.

The path is now complete. Next the logic can be developed and a tag to receive the data from the remote tag can be entered.

MESSAGE COMMAND AND LOGIC

A MSG instruction is shown in Figure D-15. A tag name must be entered for a control tag for the instruction. In this example the tag is named MSG_Control_Tag and its type is Control. When this instruction is true, it will read the specified tag in the remote controller and put it in the local tag that will be specified.

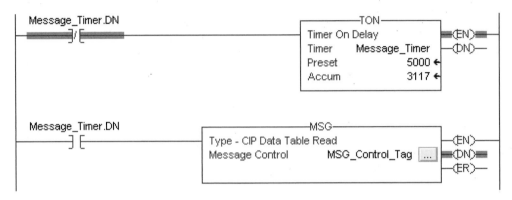

Figure D-15 Logic for the MSG instruction.

Click on the ellipsis on the MSG instruction to configure it. The screen shown in Figure D-16 will appear. CIP Data Table Read was chosen for the type. The Source Element is the name of the tag in the remote controller. The Number Of Elements to be read is 10 as the tag is an array tag with 10 members. The Destination Element is the name of the tag to which the tag data from the remote controller should be written (Temp_In[0] in this example). The table in Figure D-17 shows the possible choices for the MSG instruction.

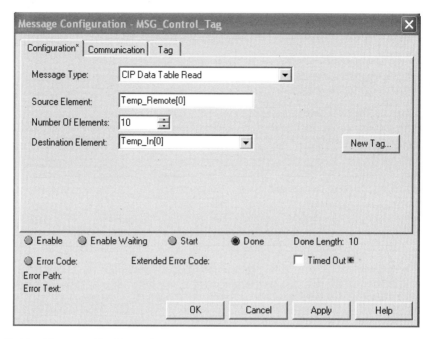

Figure D-16　Message Configuration screen.

Target Device to Communicate With	Select
Logix5000 Controller	CIP Data Table Read
	CIP Data Table Write
I/O Module that you configure using RSLogix 5000 software	Module Reconfigure
	CIP Generic
PLC-5 Controller	PLC5 Typed Read
	PLC5 Typed Write
	PLC5 Word Range Read
	PLC5 Word Range Write
SLC Controller	SLC Typed Read
MicroLogix Controller	SLC Typed Write
Block Transfer Module	Block-Transfer Read
	Block-Transfer Write
PLC-3 Processor	PLC3 Typed Read
	PLC3 Typed Write
	PLC3 Word Range Read
	PLC3 Word Range Write
PLC-2 Processor	PLC2 Unprotected Read
	PLC2 Unprotected Write

Figure D-17　Message Type choices.

Remote_CPU was entered for the path in Figure D-18. This is the name of the controller in the remote chassis (see Figure D-14).

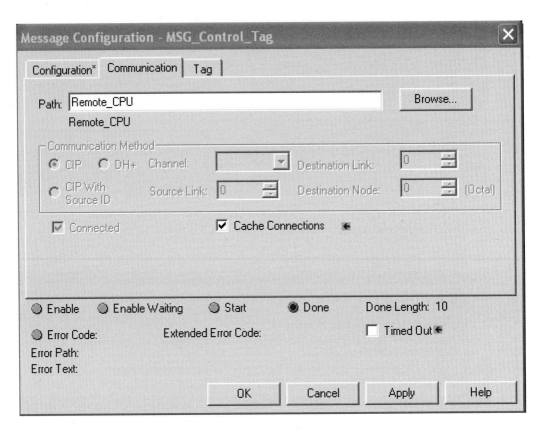

Figure D-18 Message Configuration, Communication screen.

The Tag tab was then chosen in the Message Configuration screen. MSG_Control_Tag was entered for the name of the control tag (see Figure D-19).

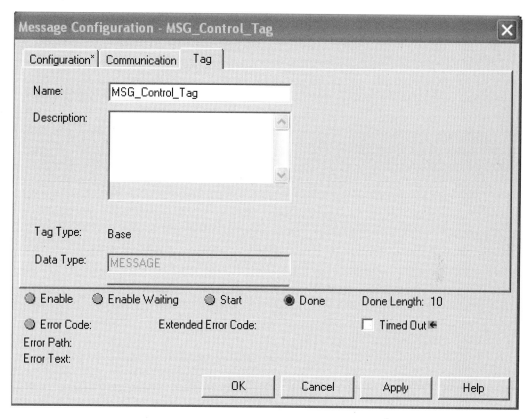

Figure D-19 Message Configuration, Tag screen.

CREATING THE TAG IN THE CONTROLLER

A controller Scope tag was created in the tag editor (see Figure D-20). The tag name is Temp_In. This is the tag to receive the data read with the MSG instruction. Note that the tag is an array of 10 members of Data Type DINT.

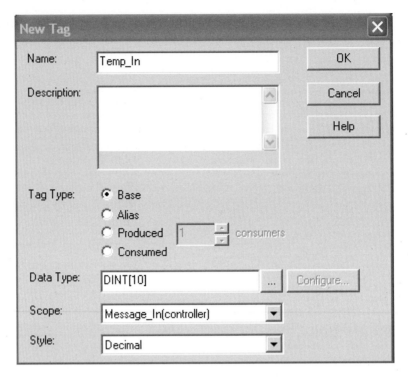

Figure D-20 New Tag configuration screen.

MSG LOGIC

Next the logic is created. Figure D-21 shows the logic that was used for this example. In this example the timer DN bit is used to execute the instruction once every 5 seconds.

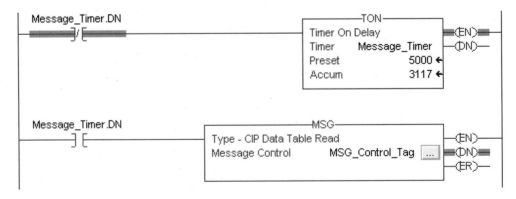

Figure D-21 Logic for the MSG routine.

Figure D-22 shows the tag array in the remote controller. Note the numbers in the 10 tag members.

Scope:	Message_Remote(c ▼	Show:	
	Tag Name △	Value	←
	─ Temp_Remote	{ . . . }	
	+ Temp_Remote[0]	220	
	+ Temp_Remote[1]	210	
	+ Temp_Remote[2]	200	
	+ Temp_Remote[3]	190	
	+ Temp_Remote[4]	180	
	+ Temp_Remote[5]	170	
	+ Temp_Remote[6]	160	
	+ Temp_Remote[7]	150	
	+ Temp_Remote[8]	140	
	+ Temp_Remote[9]	130	

Figure D-22   Remote tag in the tag editor.

Figure D-23 shows the tag array in the controller that is running the MSG instruction. Note that the numbers that the instruction read from the remote tag members are written in the tag members' Value in this controller.

Scope:	Message_In(controll ▼	Show:	S
	Tag Name △	Value	←
	+ Ethernet_Remote:I	{ . . . }	
	+ Ethernet_Remote:O	{ . . . }	
	+ MSG_Control_Tag	{ . . . }	
▶	─ Temp_In	{ . . . }	
	+ Temp_In[0]	220	
	+ Temp_In[1]	210	
	+ Temp_In[2]	200	
	+ Temp_In[3]	190	
	+ Temp_In[4]	180	
	+ Temp_In[5]	170	
	+ Temp_In[6]	160	
	+ Temp_In[7]	150	
	+ Temp_In[8]	140	
	+ Temp_In[9]	130	

Figure D-23   Tag in the tag editor in the controller executing the MSG Read instruction.

APPENDIX

E

Configuring ControlLogix

This information is written to help configure a motion project for RSLogix 5000. The objective is to correctly configure all the parameters for the drives and motors when a new motion project is developed. The configuration of the drives and motors should all be correct before beginning to develop the logic.

Start a new ControlLogix project. Click on Controller Properties, click on Properties, click on the Date/Time tab (see Figure E-1), click on Make this controller the Coordinated

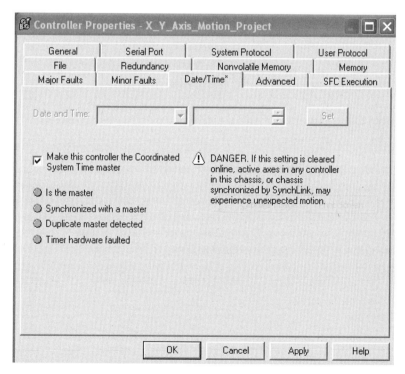

Figure E-1 Configuring the controller to be the Coordinated System Time (CST) master.

System Time master, and click OK. This makes sure that this CPU will coordinate the motion of all the axes.

Next we need to add the SERCOS card and the drives. In this example there are two drives that control two axes of motion. Figure E-2 shows the controller organizer before the modules have been added.

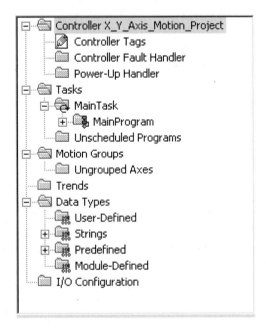

Figure E-2 Controller Organizer before any modules have been added.

Here is a quick overview of what needs to be done.

I/O modules need to be added to the project as well as the SERCOS module. Then the axes have to be configured to operate correctly in the application.

First add the SERCOS module. Right click on I/O Configuration in the Controller Organizer. Choose New Module, and the screen shown in Figure E-3 will appear. In this example a 1756-M08SE module was chosen. Select OK.

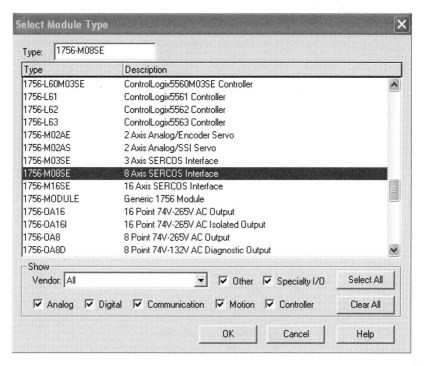

Figure E-3 Select Module Type screen. Note that a 1756-M08SE module was chosen.

Next the Module Properties screen will appear. SERCOS_Module was entered for the name. A 4 was entered for the Slot of the module. The Revision level was also entered. Compatible was chosen for the Electronic Keying method. Then Finish was selected.

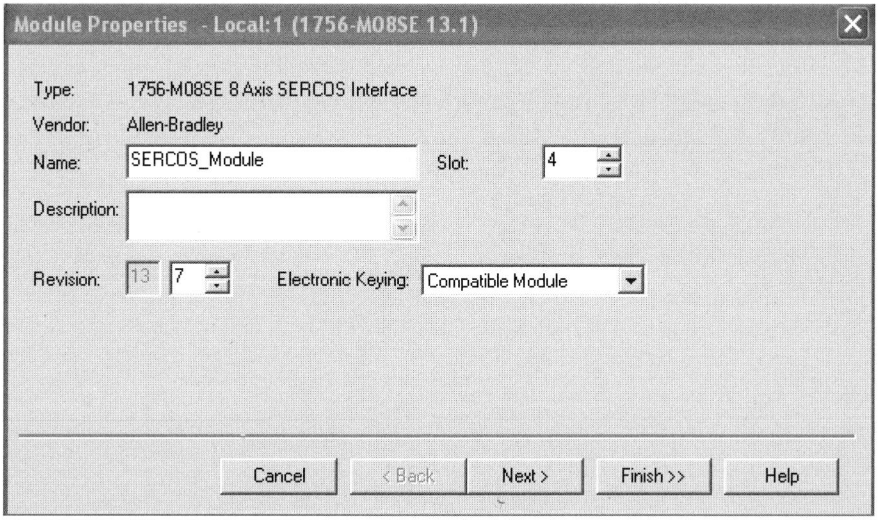

Figure E-4 Module Properties screen.

The SERCOS module is now shown under the I/O Configuration folder in the Controller Organizer in Figure E-5.

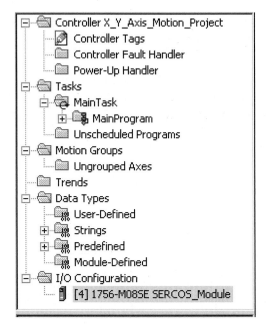

Figure E-5 Controller Organizer screen after the SERCOS module was added.

Drives must be added under the SERCOS module. The drives must be named and configured. A tag must also be created for each axis. Right-click on the SERCOS module that was just added in the Controller Organizer and choose New Module. Figure E-6 shows that a 2098-DSD-010-SE drive was selected for this application. The correct model number for the drive can be found on the drive. After OK is selected, the screen shown in Figure E-7 will appear.

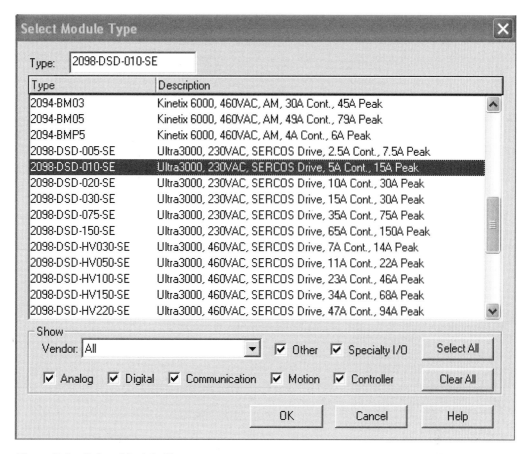

Figure E-6   Select Module Type screen.

Figure E-7 shows how the drive is configured. This drive was named X_Axis and 11 was entered for its Node address. The Node address is set on the front of the drive for this model on two dials. The Node address must be set to the same Node address on the actual drive. The drives are addressed to 11 (X_Axis) and 22 (Y_Axis). The correct Revision level was chosen and Compatible Module was chosen for the Electronic Keying method. The Next button was chosen, and the screen shown in Figure E-8 appeared.

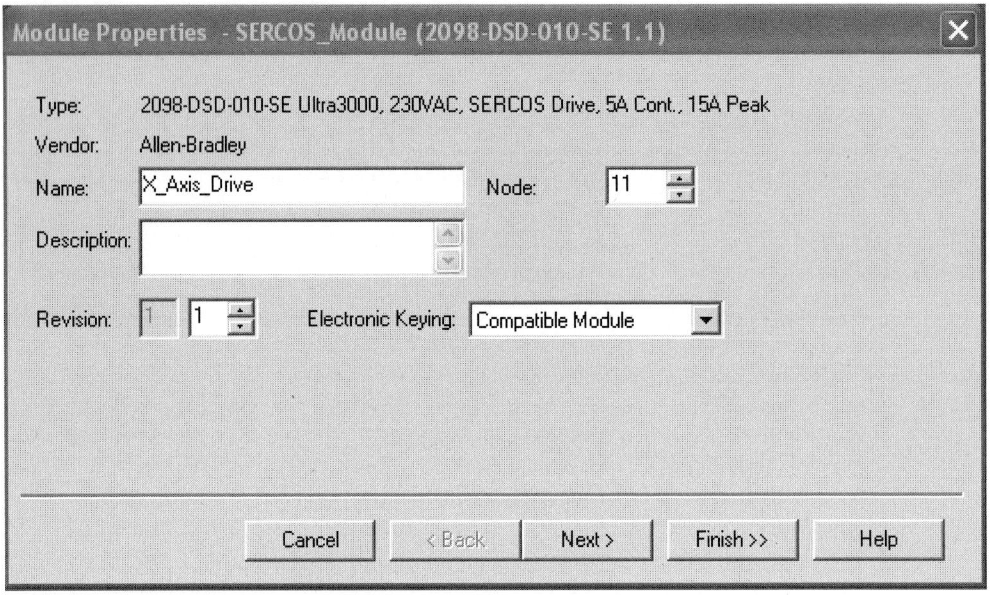

Figure E-7 Drive Name and Node number configuration.

In the screen shown in Figure E-8 you must create a New Axis tag as there is none in the selection list that have been previously created. Click on New Axis and the screen in Figure E-9 appears.

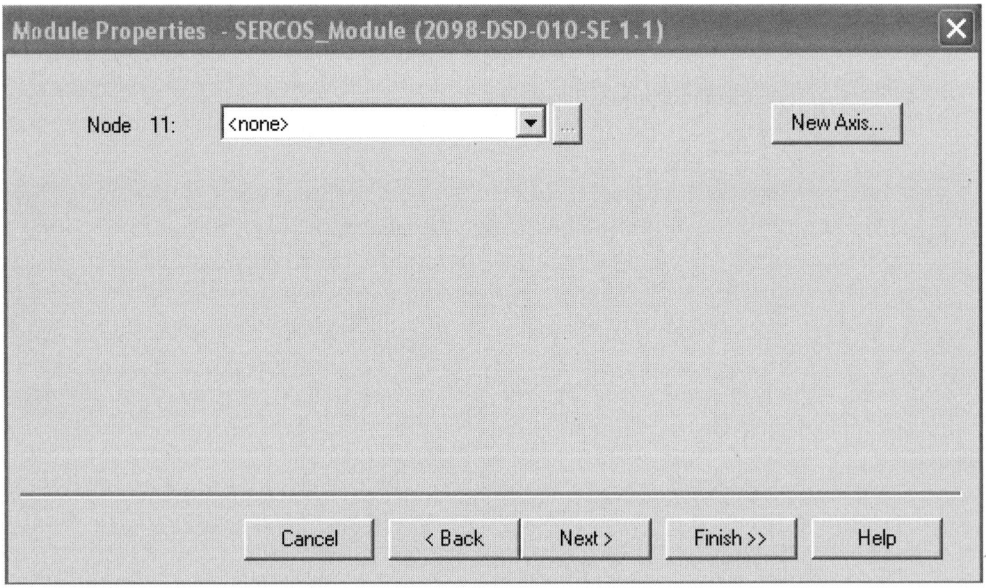

Figure E-8 Module Properties tag selection screen.

The tag name for the X_Axis is entered as X_Axis in the screen shown in Figure E-9. Its data type is Axis_Servo_Drive. OK is then selected, and the screen in Figure E-10 appears.

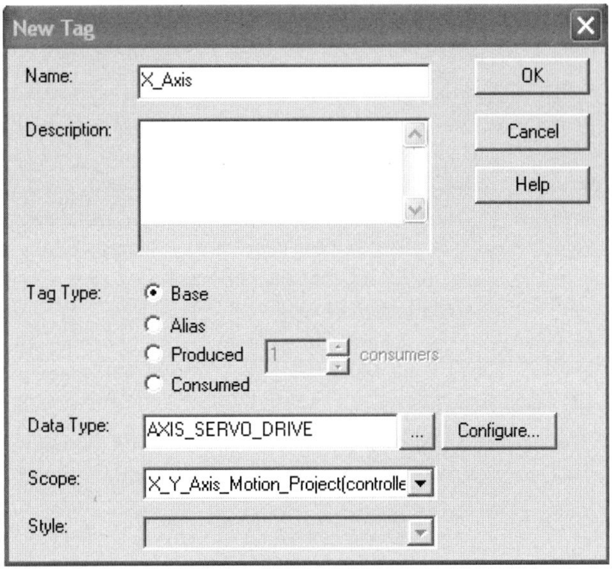

Figure E-9 New Tag screen.

Figure E-10 shows that the New Tag X_Axis was chosen for the tag for this axis. Then Finish was selected.

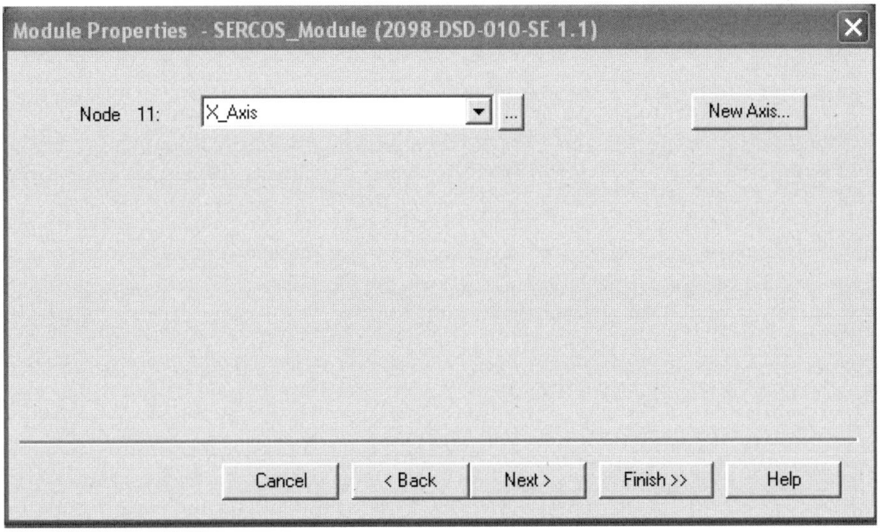

Figure E-10 Module Properties screen.

Figure E-11 shows the Controller Organizer after the first drive and its tag was added. Note the X_Axis tag under the Ungrouped Axes folder and the X_Axis_Drive under the SERCOS_Module.

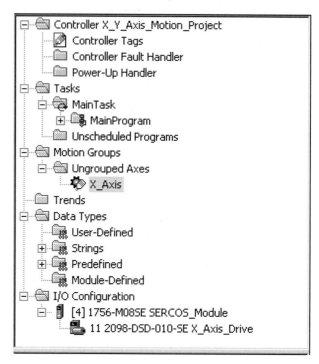

Figure E-11   Controller Organizer screen after the drive and tag were added for the X_Axis.

Next you must add the second drive. Right-click on the SERCOS module in the controller organizer and choose New Module. Choose the correct drive from the list in Figure E-12. A 2098-DSD-010-SE drive was selected for this application.

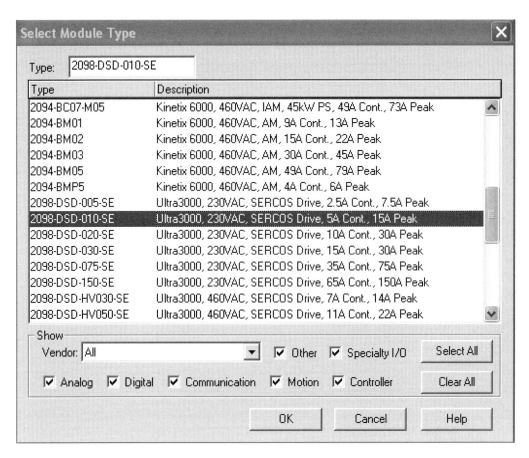

Figure E-12   Select Module Type screen.

Figure E-13 shows how this drive is configured. This drive was named Y_Axis and 22 was entered for its Node address. The Node address must be set to the same Node address on the actual drive. The correct Revision level was chosen and Compatible Module was chosen for the Electronic Keying method. Finish was chosen, and the screen shown in Figure E-14 appeared.

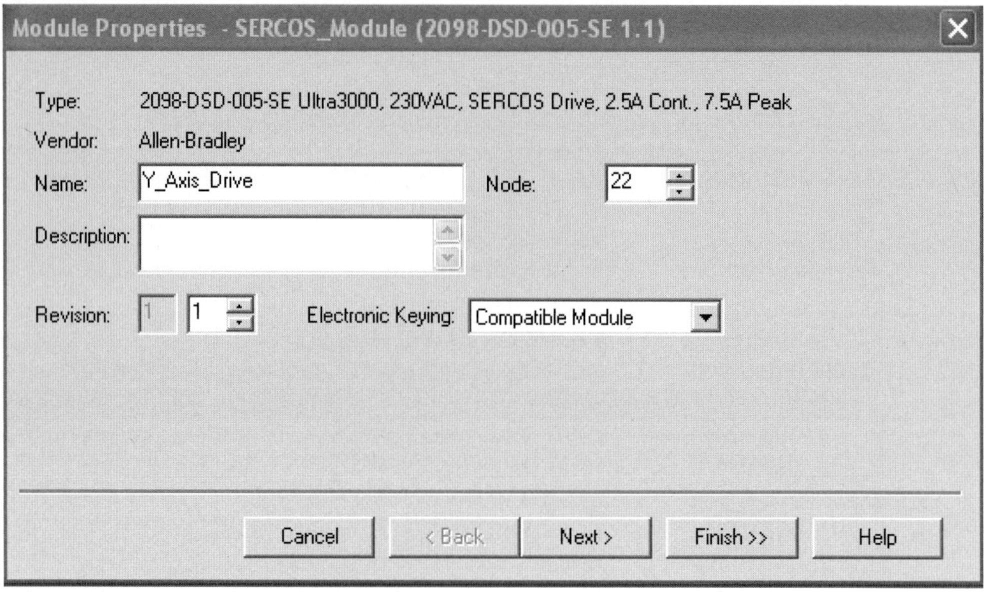

Figure E-13   Module Properties screen for the Y_Axis drive.

Next a tag must be created for the Y_Axis. The New Axis button was chosen and the screen in Figure E-15 appeared.

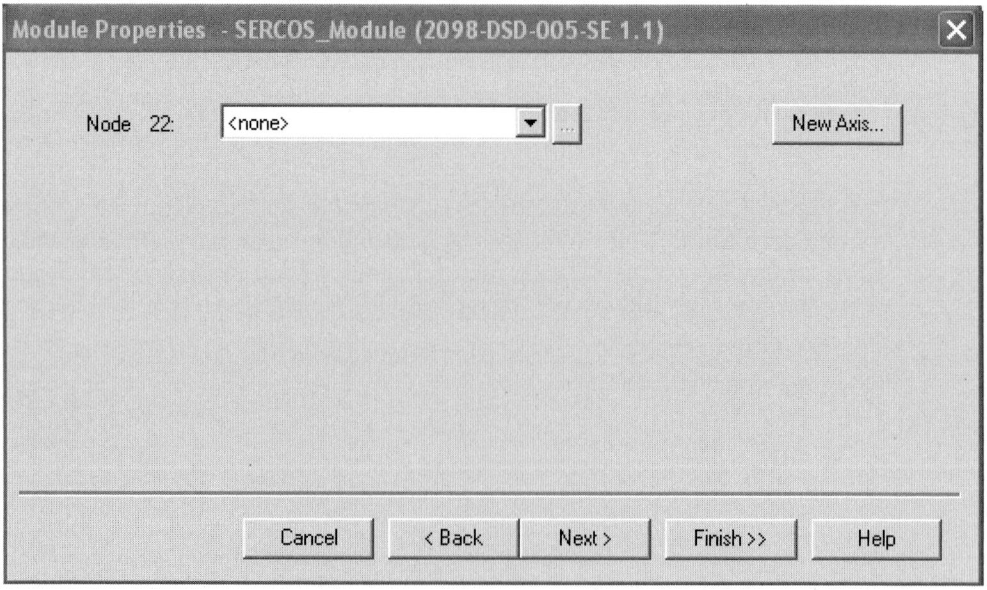

Figure E-14   Module Properties screen for the drive.

Y_Axis was entered for the Y_Axis tag name. The Data Type for the tag is AXIS_SERVO_DRIVE. Then OK was selected.

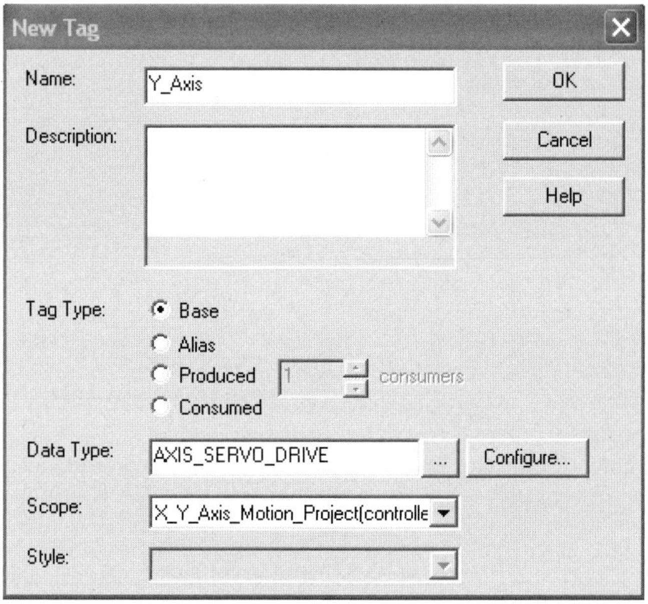

Figure E-15   New Tag screen for the axis drive configuration.

The Y_Axis tag was selected for the axis in Figure E-16. Then Finish was chosen.

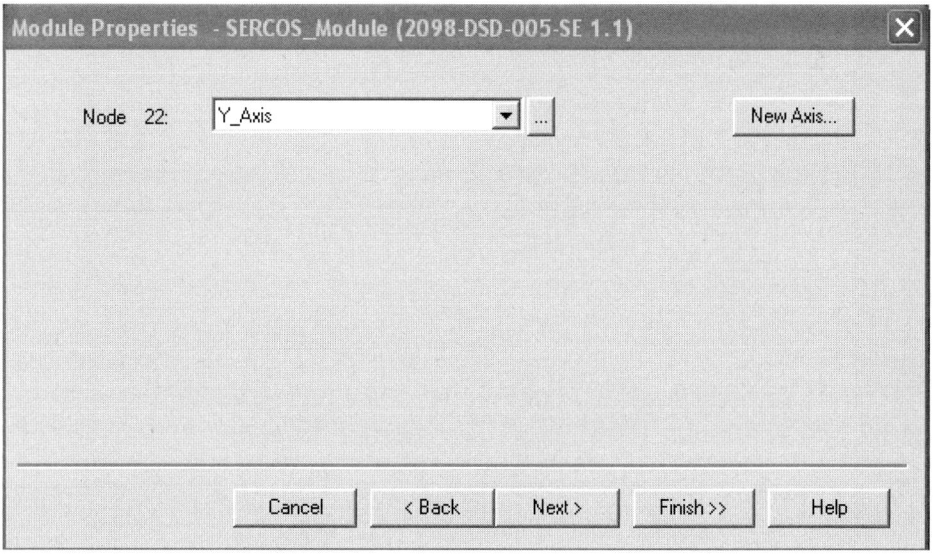

Figure E-16   Module Properties screen for the Y_Axis drive tag.

Figure E-17 shows what the Controller Organizer looks like at this point.

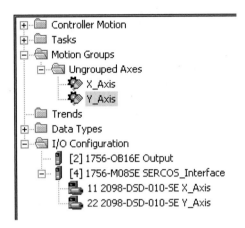

Figure E-17 Controller Organizer after the drives were added under the SERCOS module and after the tags were created. Note the tags under the Ungrouped Axis folder.

Next you need to create a Motion Groups tag. Right-click on the Motion Groups icon in the Controller Organizer and add a new group. The screen shown in Figure E-18 should appear. The motion group tag was named X_Y_Motion_Group.

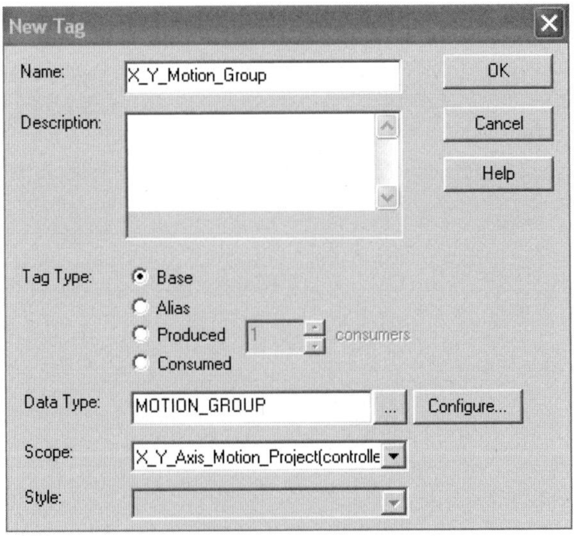

Figure E-18 New Tag screen for creating a new motion group.

Right-click on the new motion group you just created and choose Properties. The screen shown in Figure E-19 should appear.

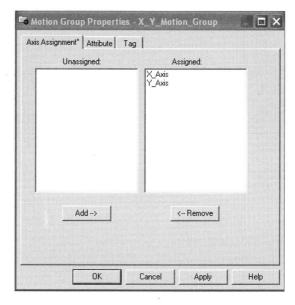

Figure E-19   Motion Group Properties screen.

Next click on the Attribute tab at the top of the Motion Group Properties screen and the screen shown in Figure E-20 will appear. Set the Coarse Update Period to 4. This number represents the number of 0.5 ms used to update the motion. You should have a minimum of 2 per axis used. In this application you have 2 axes so the Coarse Update Period should be set to 4 minimum.

Figure E-20   Motion Group Properties screen. This is where the Coarse Update Period is set. Then select Apply and OK.

Figure E-21 shows what the Controller Organizer looks like after the motion group X_Y_Motion_Group was added.

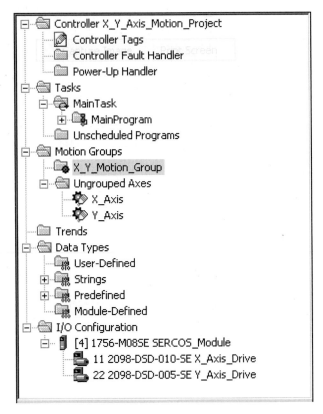

Figure E-21 The new motion group named X_Y_Motion_Group under Motion Groups in the Controller Organizer.

Next you need to move the axis tags you created into the new motion group. Left-click and drag the axes tags (X_Axis and Y_Axis) you created (under the Ungrouped Axes tag) and drop them into the motion group folder you created for this application (see Figure E-23). In this example the name of the motion group folder is X_Y_Motion_Group. The Controller Organizer is shown in Figure E-22.

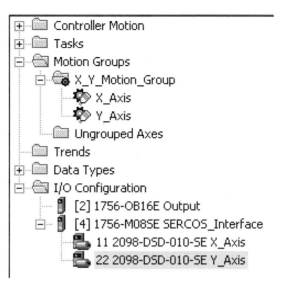

Figure E-22 Controller Organizer screen after drives, motion tags, and a motion group folder X_Y_Motion_Group were added. Note that the axes tags were dragged to the new motion group folder X_Y_Motion_Group.

CONFIGURING THE AXES OF MOTION

Next the axes need to be configured. This is done in the tag for each axis. Right-click on the X_Axis tag under Motion Group and select Axis Properties, and the screen shown in Figure E-23 will appear.

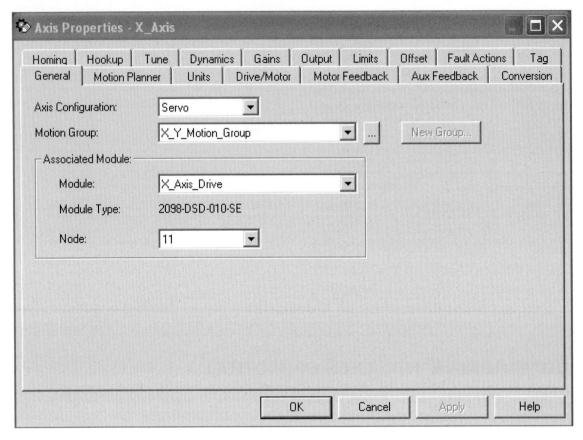

Figure E-23 Axis Properties configuration screen.

Click on the Drive/Motor tab and the screen in Figure E-24 will appear. The drive in this example is a 2098-DSD-010-SE model. The drive model number was entered into the Amplifier Catalog Number. Next the motor model is entered. Click on the Change Catalog button and select the correct motor model. The Motor Catalog Number is found on the motor. The motor model in this example was an N-3412-2-H.

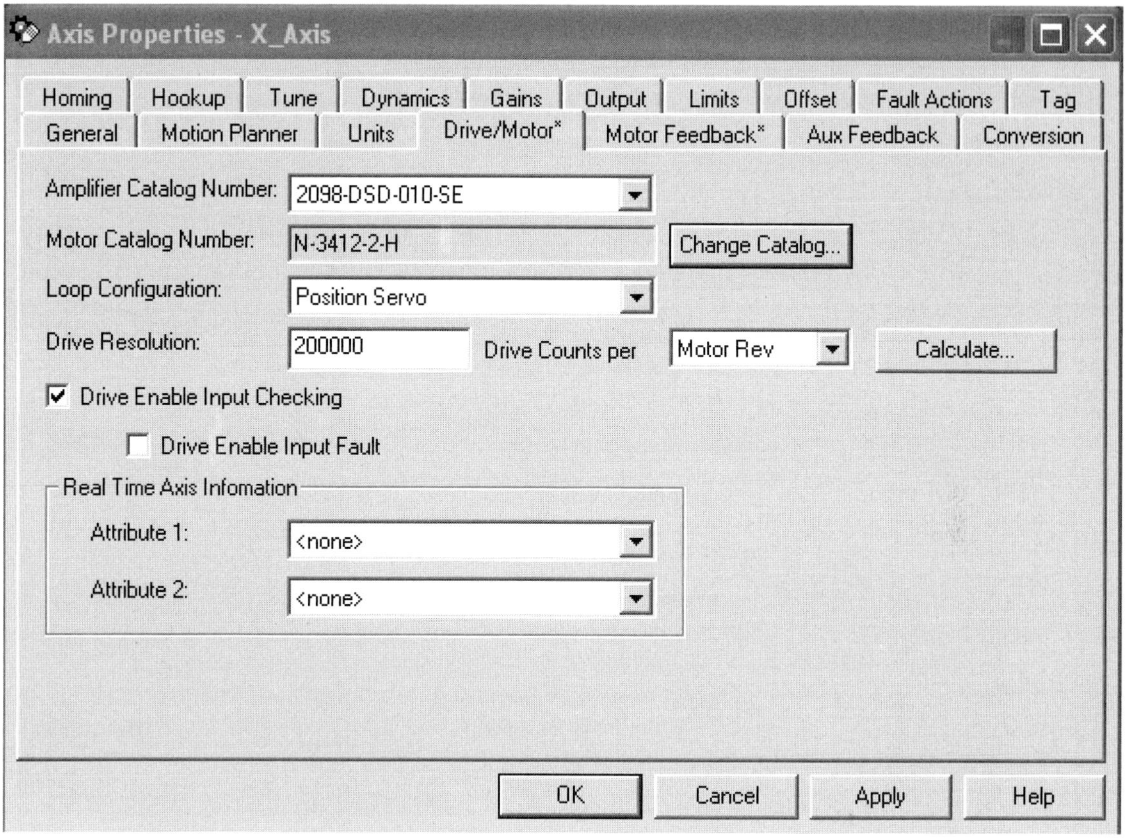

Figure E-24 Drive/Motor selection.

There are several more parameters that need to be configured to prevent damage to the drive or motor. You will be configuring several important parameters for how the drives will function. The most important to prevent damage to the axis in this application is the parameter that sets hard limits for the application. In this application if hard limits are not set, a crash could occur and do great damage. To set hard limits, click the Limits tab and check the box next to Hard Travel Limits (see Figure E-25).

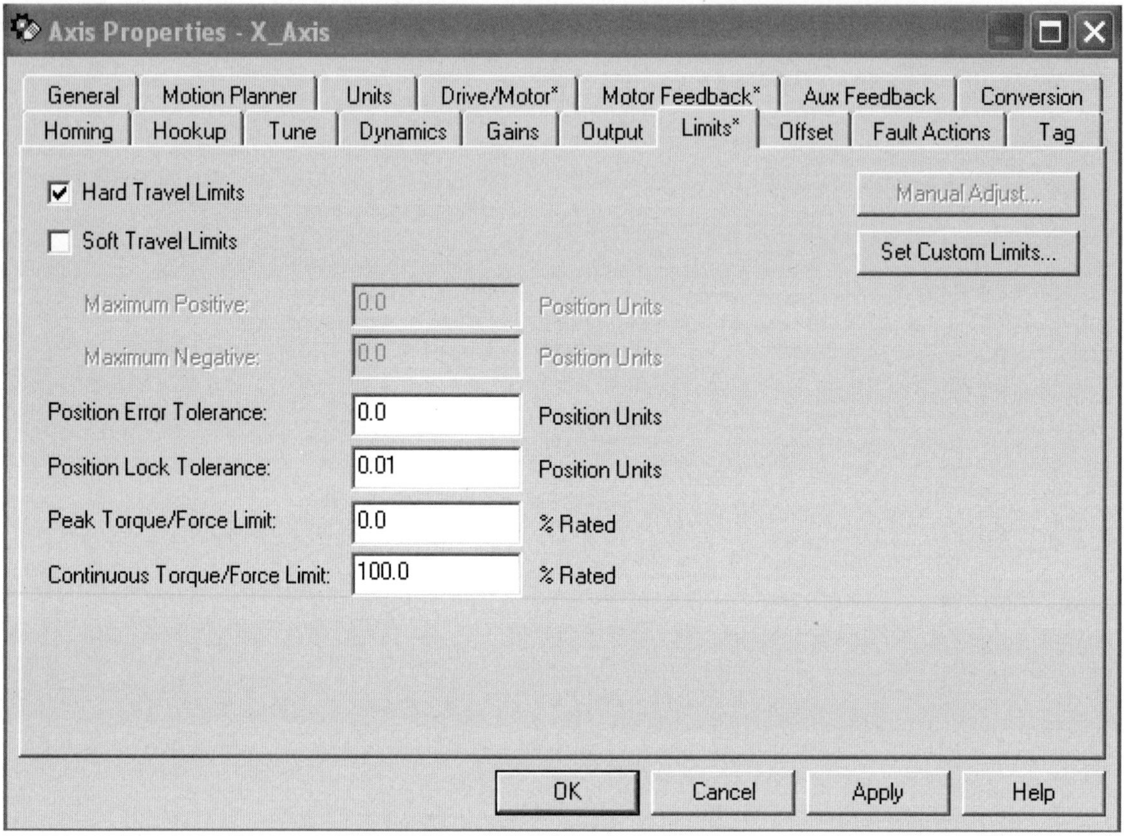

Figure E-25 How Hard Travel Limits are set for a drive.

Next choose the Fault Actions tab and select Stop Motion in the Hard Overtravel checkbox (see Figure E-26). This tells the system to stop motion if an overtravel switch goes false. Note that the end limit switches on this system are normally closed switches. This must be done individually for each axis.

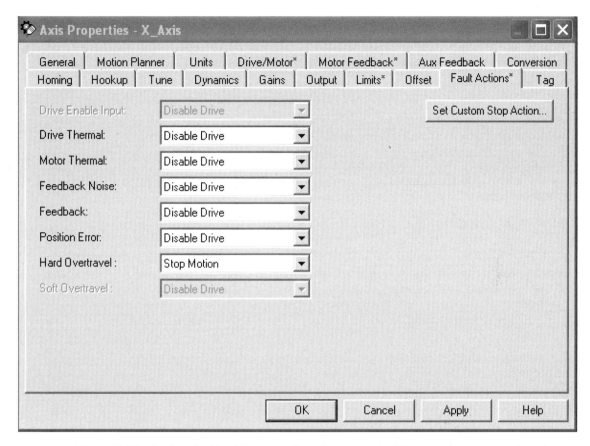

Figure E-26 Setting the Hard Overtravel configuration. In this example you want the drive to stop the motion if the axis moves too far and hits an end limit switch.

Next homing will be configured. This is configured for each axis in the axis tag. Choose the Homing tab in the Axis Properties screen (see Figure E-27). You will utilize the homing method that uses the home sensor (switch) and the index pulse on the encoder (marker pulse). Choose Switch/Marker for the type of homing, set the Limit Switch type to Normally Open, and set the homing Direction to Reverse Bi-directional. This will tell the drive that when a home command is executed it should move in a negative direction to find the home switch. The home switches in this example system are on the negative end of each axis. Speed and Return Speed homing velocities must be entered. Homing is normally done at a low speed. A 5 was entered in each for this example; 5 represents 5 inches per minute in this application.

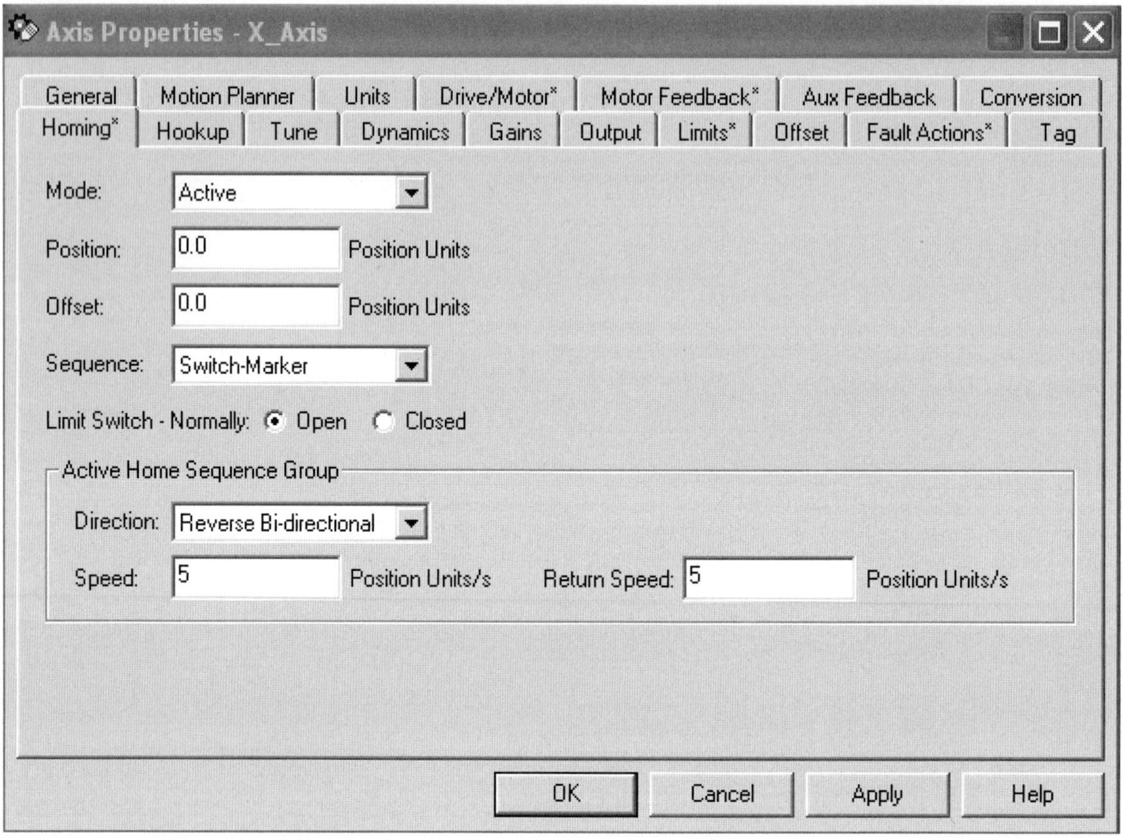

Figure E-27   Homing configuration screen.

RESOLUTION OF THE AXES

Figure E-28 shows the Axis Properties screen for the X_Axis. Note that Conversion Constant 200000 was entered in this example. This number represents Drive Counts/1.0 Position Units. In this example the desired unit of measurement for the system is inches, so 200000 Drive Counts is equal to 1 inch of travel.

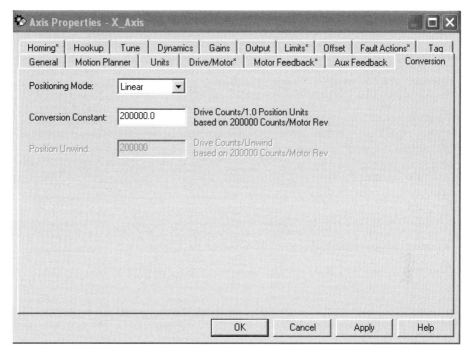

Figure E-28   Axis Properties screen for X_Axis.

A discrete output module will be required for this application. Outputs from the output module will be used to enable each drive. Right-click on I/O Configuration and choose New Module. Add an output module. Figure E-29 shows the Controller Organizer after the output module was added. Note that you could have added the output module before you added the SERCOS card and drives.

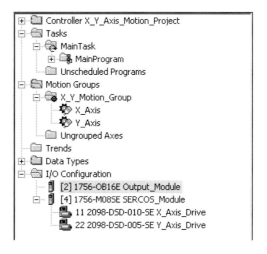

Figure E-29   Controller Organizer after the output module was added.

At this point the configuration is complete. Next you need a simple program to enable the drives. In this example the drive enable input for the X_Drive was connected to output 0 on the output module in slot 2. The enable input for the Y_Drive was connected to output 1 on the output module in slot 2. Figure E-31 shows a simple ladder diagram that could be used to enable the drives.

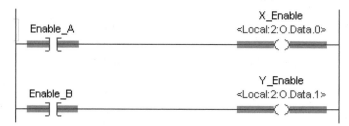

Figure E-30   Logic to enable the two drives. Note that the outputs from the output module are connected to the drive enable input on each drive.

At this point logic can be written to command motion or the drives can be tested with motion direct commands.

MOTION DIRECT COMMANDS

Motion direct commands can be used to test your axes before you write the logic. You must be online in order for any of these commands to work. Right-click the axis icon for the axis you want to test, then click on Motion Direct Commands. The Motion Direct Commands window opens; first highlight MSO, the Motion Servo On instruction, and then hit Execute. The MSO instruction closes the servo loop and puts the drive in control. From here there are several commands that can be used.

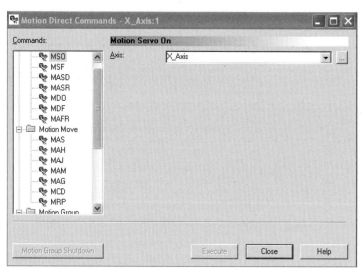

Figure E-31   Motion Direct Commands screen.

First the drives must be enabled. Figure E-30 shows a simple ladder diagram that turns on the drive enable input for each axis. A BOOL tag was used to control the output to each drive's enable. This is done to electrically enable and disable each drive.

If the CLX is put in Run mode and the BOOL tags are energized to enable the drives, motion direct commands can be used to test the axes.

Right-click on one of the drives and choose Motion Direct Commands.

The first command that must be executed is an MSO. This closes the servo loop for the drive. This must be done, or the drive will not execute any commands.

When an axis is first powered up, it does not know its current position. It needs to be *homed* to establish its current position. The motion direct command MAH will use the parameters that were set up above in the homing properties. Figure E-32 shows the Motion Direct Commands screen and the MAH command in the list. Note that the X_Axis was chosen in this example. If the Execute button is chosen, the drive should initiate the homing routine.

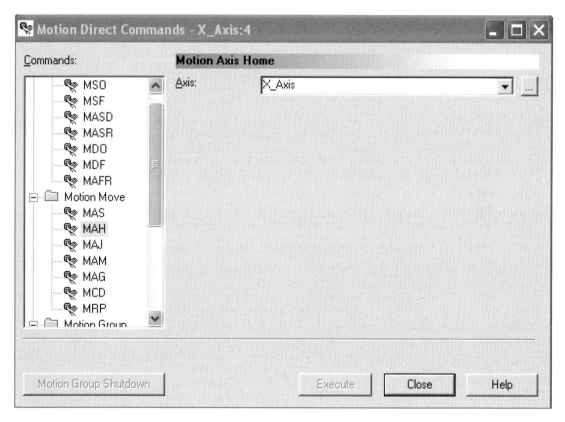

Figure E-32   Motion Direct Commands screen.

Once the axis is homed, other commands may be tried. A motion axis jog (MAJ) instruction is used to jog an axis. With a MAJ command you have to use a motion axis stop (MAS) command to stop the axis. When the MAJ command is executed in incremental mode, the drive will continue to move until a MAS command is executed for that axis. To use an MAJ instruction, you must choose a Direction and enter a Speed. Forward was chosen for direction and 2 was entered for the MAJ motion direct command in Figure E-33. Note the X_Axis was chosen for the Axis. When the Execute button is chosen, the axis will move in a positive direction until a limit is encountered or a MAS instruction is executed.

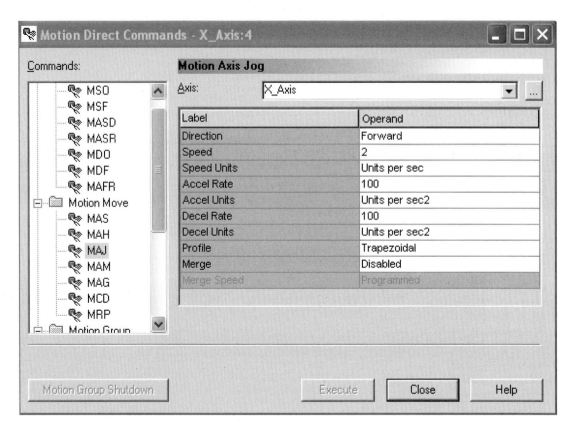

Figure E-33   MAJ command.

A MAS instruction is shown in Figure E-34. Note that X_Axis was chosen. If the Execute button is chosen, the MAS command will stop motion of the X_Axis.

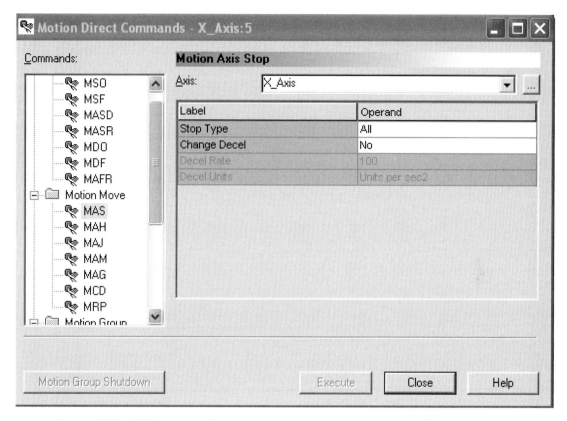

Figure E-34   MAS command.

There are many other motion direct commands that can be executed. The use of motion direct commands can help understand motion programming instructions available in CLX.

INDEX